The Windmill: Its Effeciency and Economic Use

by Edward Charles Murphy

with an introduction by Roger Chambers

Self Reliance Books

Get more historic titles on animal and stock breeding, gardening and old fashioned skills by visiting us at:

http://selfreliancebooks.blogspot.com/

Introduction

I am pleased to present yet another title on Homesteading and Farm Life.

This volume is entitled "The Windmill: Its Effeciency and Economic Use " and was published in 1901.

The work is in the Public Domain and is re-printed here in accordance with Federal Laws.

As with all reprinted books of this age that are intended to perfectly reproduce the original edition, considerable pains and effort had to be undertaken to correct fading and sometimes outright damage to existing proofs of this title. At times, this task is quite monumental, requiring an almost total "rebuilding" of some pages from digital proofs of multiple copies. Despite this, imperfections still sometimes exist in the final proof and may detract from the visual appearance of the text.

I hope you enjoy reading this book as much as I enjoyed making it available to readers again.

Roger Chambers

Fig. 11—C. E. Colburn's Farm and Stock Barn

1

Fig. 19—MR. LAWSON VALENTINE'S BARN, "HOUGHTON FARM," MOUNTAINVILLE, N. Y.

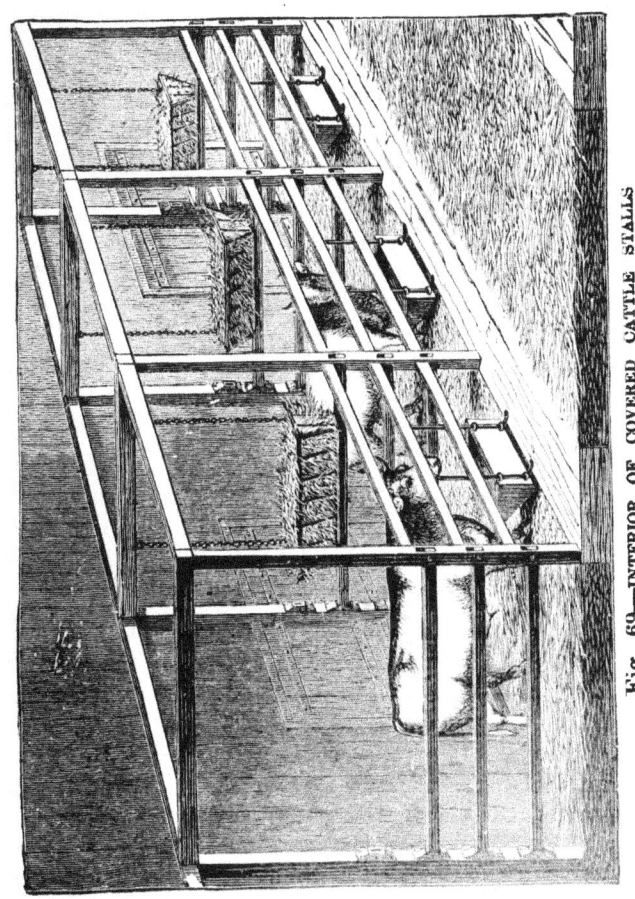

Fig. 69—INTERIOR OF COVERED CATTLE STALLS

3

Fig. 92—NOVA SCOTIA DAIRY AND FRUIT BARN

4

CONTENTS.

ILLUSTRATIONS.

A. VIEW OF MILL NO. 20—15½-FOOT JUMBO.

B. VIEW OF DEFENDER MILL AND PUMP KNOWN AS "WATER ELEVATOR."

A. VIEW OF MILL NO. 57—24-FOOT LITTLE GIANT.

B. VIEW OF MILL NO. 55—7½-FOOT JUMBO.

THE WINDMILL: ITS EFFICIENCY AND ECONOMIC USE.

PART I.

By Edward Charles Murphy.

INTRODUCTION.

History does not record the name of the person who invented the windmill, nor give the date of the invention. The belief that windmills were used by the Romans is not well authenticated, and their use by the Bohemians in 718 is doubted. It is quite clear, however, that they were used in France and Italy in the twelfth century for grinding corn, and that they were used in Holland in the fifteenth century for pumping water over the dikes into the sea.

Mr. John Burnham, of Connecticut, is said to be the inventor of the American windmill. Mr. L. H. Wheeler, an Indian missionary, patented the Eclipse mill in 1867. The first steel mill was the Aermotor, invented by Mr. T. O. Perry in 1883.

The common European windmill, shown in section in figs. 1 and 2, differs much in appearance from the American mill. The wind wheel of the European mill has usually four long wooden arms, to each of which is attached a sail, against which the wind presses. The sails consist of a framework, on which canvas is stretched, usually forming a warped surface, the angle with the plane of the arms (called the angle of weather) being about 7° at the outer end and 18° at the inner end. The length of sail was usually about five-sixths the length of the arm, the width of the outer end one-third the length, and the width of the inner end one-fifth the length. The sail area is seen to be small compared with the wind area or zone containing the sails. The arms were sometimes 60 feet long. The American wooden mill is much smaller and more compact than the European mill. It has six or eight arms, to which is attached a framework carrying many small sails. These sails are usually 3 or 4 feet long, 3 or 4 inches wide at the outer end, and 1 to 3 inches wide at the inner end, and are set at an angle of 30° to 40° to the plane of the wheel. For large wheels these sails are arranged in two or more concentric rings. The American steel mill differs principally from the wooden mill in that it has larger and fewer sails for a given size of mill, its sails are curved instead of plane, and it offers less resistance to the passage of the air over the back of the sails.   In the sails of the steel mill there is seen to be a partial return to those of the European mill.

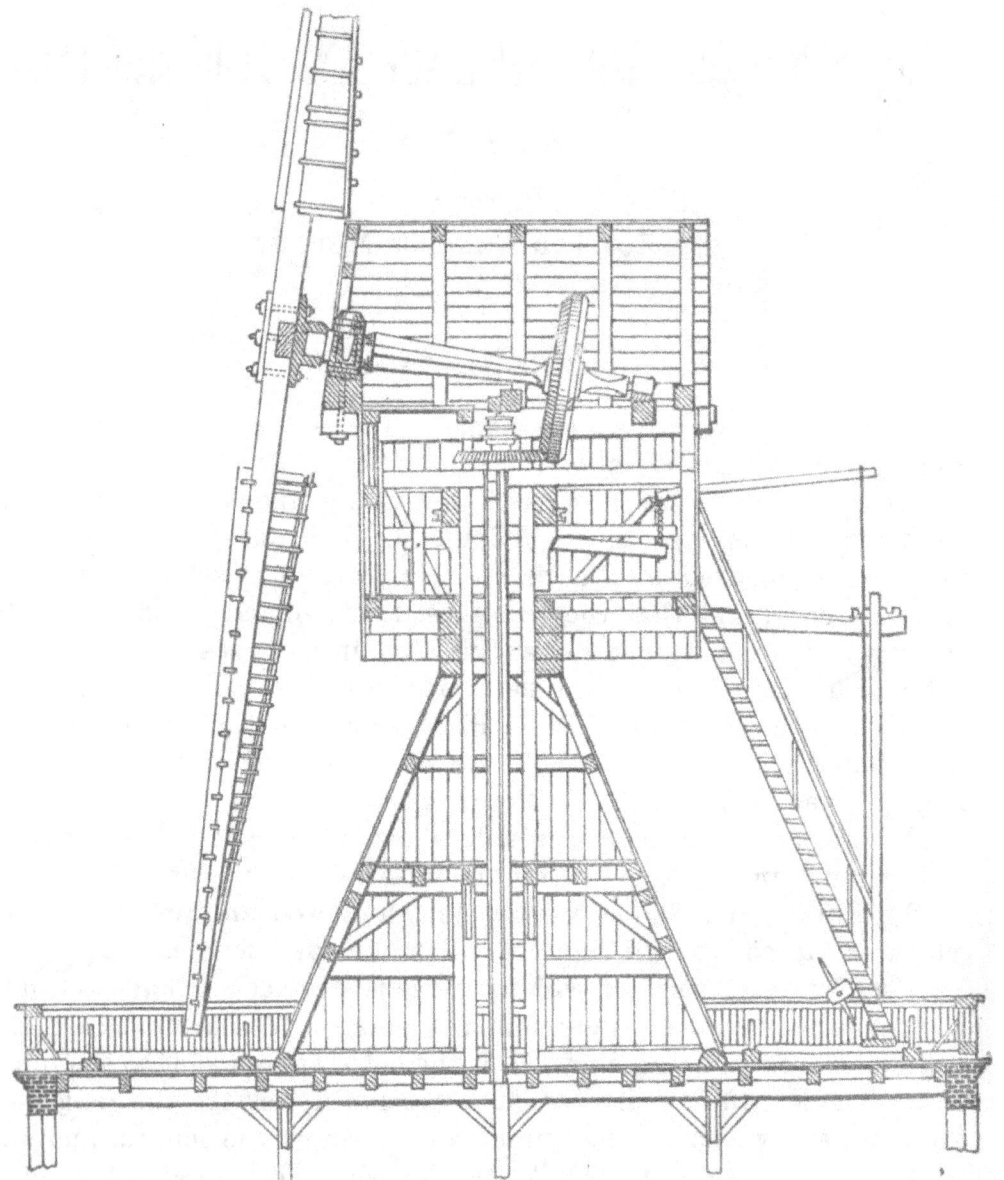

FIG. 1.—Section of common European post windmill mounted on central column.

CLASSIFICATION OF WINDMILLS.

Windmills may be divided into two general classes—paddle wheel and sail wheel.   The Jumbo shown in Pl. XVI, *B* (Part II), and Little Giant No. 56, described on pages 125 to 127, Part II, are good illustrations of the first class.   In both of these mills the sails move with the wind, and it is necessary to have a shield, or a method of feathering the sails, in order to keep the wind from striking them when they are moving in a direction opposite to that of the wind. In the Jumbo the axis of the wheel is horizontal, in the Little Giant it is vertical; but the wind acts on the sails of both in substantially the same way. The air acts with full pressure on only one sail of the Jumbo at any one time, and on half the sails it has no action, or only negative action.

In the sail-wheel mill (fig. 1, Part I, and Pl. XV, Part II) the wheel moves at right angles to the direction of the wind, instead of in the same direction, as in the paddle-wheel mill. The wind acts with a

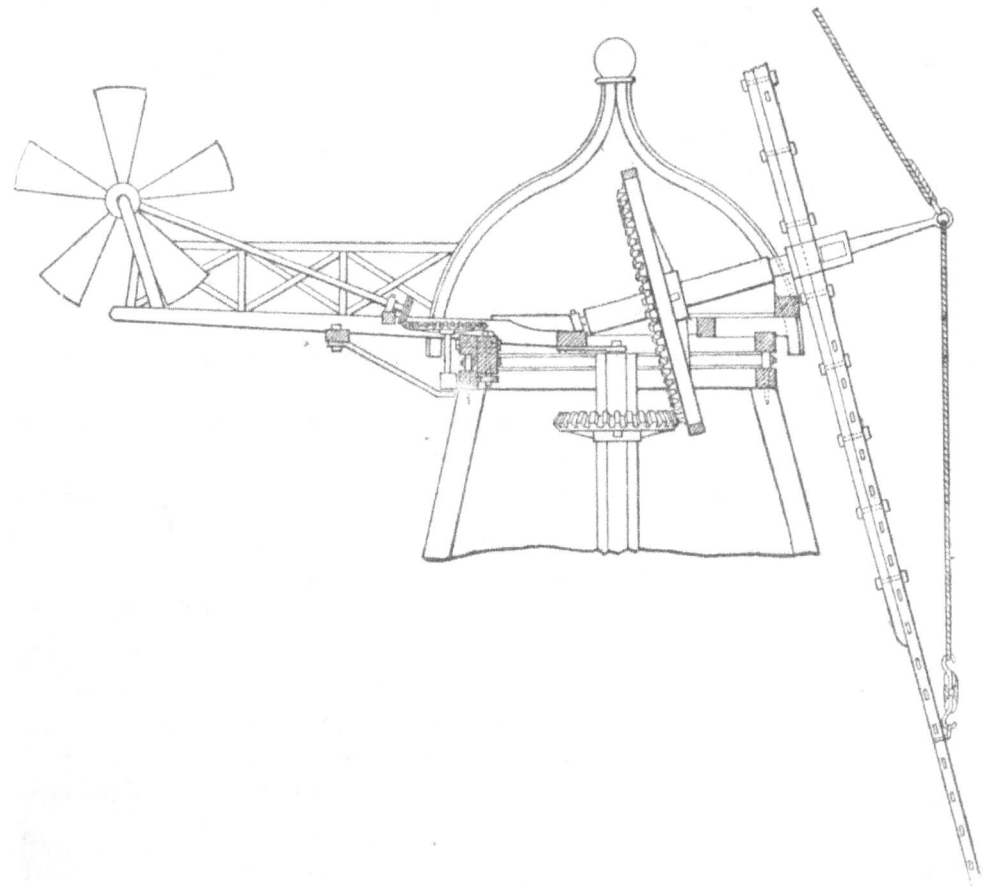

Fɪɢ. 2.—Early form of head of European tower windmill.

certain pressure on all of the sails all of the time. The circumference velocity of the sail wheel may be two or more times greater than the velocity of the wind that drives it, but the circumference velocity of the paddle wheel is always less than the velocity of the wind that drives it. The sails of the sail-wheel mill must be placed at an angle with the plane of the wheel, so that the wind will press on them; but the sails of the paddle-wheel mill may be in the plane of the axis of the wheel.

For greatest pressure on the sails of a sail-wheel mill the axis of the wheel must be parallel to the direction of the wind. It is necessary, then, for the wheel to change its direction as the wind changes. In the German or post mill the whole building, as well as the wheel, can be turned around on a post by hand. In the tower or Dutch mill the upper part only of the mill turns with the wheel. This is accomplished by hand in the mill shown in fig. 1, and by an auxiliary windmill in the mill shown in fig. 2. In the American mills the upper part only turns on a turntable, which is usually on rollers or balls. This is accomplished, first, by the pressure of the wind on a long rudder vane extending out behind the mill, and in some mills by a side vane as well; or, second, by the pressure of the wind on the wheel itself, which is placed on the opposite side of the tower, as in fig. 46, Part II; or, third, by side wheels, as in mill No. 52, fig. 50, Part II.

REGULATING DEVICES.

The wind is constantly changing in velocity as well as in direction, and if the load on the mill is constant the speed of the mill and of the machine which it operates will change with it. If the speed is to be kept nearly constant, some device is needed to reduce the wind pressure on the wheel when the wind velocity reaches a certain amount. In the European mill there are two methods of doing this, viz, by means of a brake or friction rings and by changing the sail area. The latter is accomplished by rolling or unrolling the canvas sails by hand or automatically. In the American mill the speed is regulated by changing the sail area in one of three ways: (a) By sections of the wheel revolving about an axis which places each sail at an angle to the direction of the wind less than 90°, as in fig. 36, Part II; (b) by placing the axis of the wind wheel eccentric to the axis of the tower, so that the wind pressure on the wheel will cause it to revolve around the axis of the tower; and (c) by the wheel moving naturally around the axis of the tower in the direction in which it is revolving. This turning action around the axis of the tower is counter-acted by a spring or a weight, or, as we commonly say, "the wheel is held in the wind" by a spring or a weight. A weight is better than a spring, for by moving it out or in any desired pull can be placed on the wheel. A spring can not easily be adjusted and may lose some of its tension. The first method of regulation, called the centrifugal-governor method, is used in the Halliday mill (fig. 20) and in the Althouse-Wheeler mill (fig. 36, Part II). The second method is illustrated in the Aermotor (fig. 33, Part II), which shows the axis of the wheel eccentric to the axis of the tower by 4.5 inches. It also shows the spring which holds it in the wind. This spring also resists the action of the load, which tends to turn the wind wheel out of the wind. The third method is illustrated by the Woodmanse mill.

EARLY EXPERIMENTS.

Smeaton's experiments.—The first experiments of which we have any record were made by John Smeaton, an English engineer, and published in 1755 to 1763.[1] These experiments were made on model windmills of the European type, for the purpose of determining the best shape of sail for a given sail area. The models each had four arms 21 inches long. For one set of tests the sails were 5.6 inches broad and 18 inches long, giving an area of 403 square inches. In another set of tests the sails were 18 inches long, 5.6 inches wide at the inner end and 8.4 inches wide at the outer end, with an area of 504 square inches. The sails were either plane or warped at various angles. These mills were worked by moving the windmill around in a circle of $5\frac{1}{2}$ feet radius, in still air, instead of placing them on a tower and allowing the natural wind to drive them. The wheel was moved around in this circle by means of a cord wound on a drum on a vertical shaft, the horizontal arm which held the wheel being fastened to this shaft. The work done by the wheel in a given time was measured by observing the length of string wound on the shaft of the wheel, a weight of known size being attached to the end of the string. The velocity of the wind, which was assumed to be the velocity of the end of the arm where the wheel was attached to it, varied from $4\frac{1}{3}$ to $8\frac{3}{4}$ feet a second, or from 2.9 to 6 miles an hour.

It will be noticed that these wheels are only 3.5 feet in diameter, that they were moved in a circle only 5.5 feet in diameter, and that wind velocities or wheel velocities were small and of only a limited range—from about 3 to 6 miles an hour. Smeaton draws the following conclusions from his experiments:

(1) The velocity of the windmill sails, whether unloaded, or loaded so as to produce a maximum, is nearly as the velocity of the wind: their shape and position being the same.

(2) The load at the maximum is nearly, but somewhat less than as the square of the velocity of the wind: the shape and position of the sails being the same.

(3) The effects of the same sails at a maximum are nearly but somewhat less than as the cubes of the velocity of the wind.

(4) The load of the same sails at the maximum is nearly as their squares and their effects as the cubes of their number of turns in a given time.

(5) When the sails are loaded so as to produce a maximum at a given velocity of the wind, and the velocity of the wind increases, the load remaining the same: first, the increase of effect, when the increase of the velocity of the wind is small, will be nearly as the squares of those velocities; secondly, when the velocity of the wind is doubled, the effects will be nearly as 10 to 27.5; but, thirdly, when the velocities compared are more than double of that where the given load produces a maximum, the effects increase nearly in a simple ratio of the velocity of the wind.

(6) If sails are of similar figure and position, the number of turns in a given time will be reciproca'ly as the radius or length of the sail.

---

[1] Philos. Trans. Royal Soc. London, 1755-1763.

(7) The load at a maximum that sails of a similar figure and position will overcome at a given distance from the centre of motion, will be as the cube of the radius.

(8) The effects of sails of similar figure and position are as the square of the radius.

(9) The velocity of the extremity of Dutch sails, as well as of enlarged sails, in all their usual positions, when unloaded, or loaded to a maximum, is considerably quicker than the velocity of the wind.

Regarding the ratio of the sail area to the wind area or zone containing the sails, he found that where the ratio was greater than 7 to 8 the power of the mill was decreased instead of increased. Regarding the proper shape of sail, he found that the warped sail was more effective than the plane sail. He also found the following six angles of weather at equal distances from the shaft outward advantageous: 70°, 71°, 72°, 74°, 77.5°, and 83°. He states that a difference of two or three degrees in the angles of impact makes little difference in the power of the mill.

Coulomb's experiments.—C. A. Coulomb, a French engineer, made some tests of the work done by a Dutch windmill used for extracting oil from rape seed at Lille, in Flanders. His observations were published in 1821.[1] The mill was 70.2 feet in diameter. It had four warped canvas sails, each 28.7 feet long and 6.6 feet wide; the width of canvas was 5.5 feet. The angle which the plane of sail made to the plane of the wheel varied from 30° at the inner end to 12° at the outer end. The wind velocity was measured by the use of feathers carried along by the wind. Two men were stationed 150 feet apart, on slight elevations, to note the time required for each feather to pass over this distance. The velocity of the wind striking the wind wheel was assumed to be that found from these feathers. In a 14.9-mile wind, and with the load ordinarily used, he found that the wheel made 13 revolutions per minute with all the canvas spread. From these data he figured the useful work done by the mill per minute to be 232,388 foot-pounds and the useless work expended in shock of stampers and friction to be 37,310 foot-pounds, or a total of 269,698 foot-pounds, equal to 8.17 horsepower.

Coulomb did not consider this a complete or satisfactory test of this mill. He did not control the working of the mill, but simply observed what it did when handled by the miller who extracted the oil. He tried to induce the owner to permit him to use the mill for a time for experimental purposes, but did not succeed.

It will be seen from what follows that even if we assume the wind velocity to be correctly measured, this test does not necessarily show the power of the mill, for we do not know that the load used was the proper load for the wind velocity. It shows what the mill was doing, not what it might do under a better loading.

---

[1] Theorie de Machines Simple, by C. A. Coulomb. Paris, 1821.

Griffiths's experiments.[1]—In 1891–92 Mr. J. A. Griffiths made tests of six windmills used for raising water, with the following results:

Results of windmill tests by J. A. Griffiths.

No.	Type of windmill tested.	Outer diameter of sail.	Inner diameter of sail.	Gross area of sail wheel.	Weather angle at outer end of sail.		Diameter of pump.	Stroke of pump.	Average head of water.	Load on pump, per stroke.[a]	Velocity of wind per hour.[b]	Velocity of mill per minute.[b]	Horsepower.[b]
		Feet.	*Feet.*	*Sq. ft.*	°	′	*In.*	*In.*	*Feet.*	*Ft.-lbs*	*Miles.*	*Rev.*	
1	Toowoomba.........	22.3	8.3	39.20	18	47	5	6.75 {	25	120	4.3	5.0	0.018
									100	480	7.0	6.8	0.098
2	Stover	11.5	4.5	104.0	43		3	4 {	29.2	29.2	5.8	13.0	0.011
									61.2	61.2	6.5	13.3	0.025
3	Perkins..........	16.0	6.0	201.0	36		3	10.75	39.0	105.0	6.0	7.5	0.024
4	Althouse-Wheeler .	14.2	4.5	157.0	30		3	c10	66.3	166.0	7.0	12.6	0.065
5	....do	10.2	3.8	81.0	28		3	4.625	38.7	51.0	8.5	20.5	0.028
6	Carlyle	9.8	4.2	80.0	50		3	4	30.7	30.7	6.0	12.5	0.012

a These pump loads have been computed for comparison of these results with others.
b At maximum efficiency.
c Pump is double acting; this is twice the length of stroke.

No. 1 was a 22.5-foot wooden mill, with 20 warped sails each 87 by 36 by 9 inches, the weather angle at the outer end being 18° 47′ and at the inner end 40° 20′. It worked a direct-acting single-stroke pump having a 5-inch cylinder and 6.75 inches stroke. Two lifts were used, one of 25 feet and the other of 100 feet.

No. 2 was a 12-foot Stover wooden mill, having a wind wheel somewhat like that of mill No. 38. It had 112 sails, each 43 by 3.75 by 1.5 inches, set at an angle of 43° to the plane of the wheel. It worked a direct-acting single-stroke pump having a 3-inch cylinder and 4 inches stroke. The lifts were 29.2 and 61.2 feet.

No. 3 was a 16-foot Perkins solid wood wheel, the wind wheel having 160 sails, each 60 by 4 by 1.5 inches, set at an angle of 36° to the plane of the wheel. It worked a direct-acting double-stroke pump having a 3-inch cylinder and 5.375 inches stroke. The lift was 39 feet.

No. 4 was a 14-foot Althouse-Wheeler wooden sectional mill, the wind wheel having 104 sails, each 48 by 4 by 1.5 inches, set at an angle of 30° to the plane of the wheel. It worked a direct-acting double-stroke pump having a 3-inch cylinder and 5 inches stroke. The lift was 66.3 feet.

No. 5 was a 10-foot Althouse-Wheeler wooden sectional mill. The wind wheel had 84 sails, each 38 by 3.75 by 1.5 inches, set at an angle of 28° to the plane of the wheel. It worked a direct-acting single-stroke pump having a 3-inch cylinder and 4.625 inches stroke. The lift was 38.7 feet.

No. 6 was a 10-foot Carlyle iron mill. It had 7 somewhat spoon-shaped sails, each having a spout-like extension through which the

[1] Windmills for raising water, by J. A. Griffiths: Proc. Inst. Civ. Eng., Vol. CXIX, p. 321.

air flowed. It worked a direct-acting single-stroke pump having a 3-inch cylinder and 4 inches stroke. The lift was 30.7 feet.

The wind velocity was measured with the "f" wind gage, which was either on a tower near by or on an arm projecting as far as possible to windward. It appears that it was necessary to be near the gage in order to read the velocity, which would indicate a possible error in wind velocity and some interference with the wind striking the wind wheel. The range of wind velocities is not stated, and not more than two loads were used in any case.

These results will be compared with others further on.

King's experiments.[1]—Prof. F. H. King conducted a series of experiments with a 16-foot geared Aermotor, covering a period of one year—from March 6, 1897, to March 6, 1898. This mill was used to work one or more of four pumps: (1) A reciprocating piston pump with a 14-inch cylinder and 9 inches stroke; (2) a bucket pump having a normal capacity of 120 gallons per minute; (3) a No. 2 Gould centrifugal pump; and (4) the smallest size Menge pattern centrifugal pump. The bucket pump was used nearly all of the time. The reciprocating piston pump was used occasionally by itself and part of the time with the bucket pump, when the wind velocity was strong enough to carry both. The Menge was used occasionally with the piston pump and the bucket pump, when the wind velocity was strong enough to carry all.

There was no automatic device for throwing into or out of use any of these pumps. It was necessary to do this by hand, so that a part of the time the load on the mill was not suited to the wind velocity. This can be seen from the record. For example, on February 10, from 1 to 7 p. m., the wind velocity varied from 9 to 12 miles an hour and 8.6 tankfuls of water were pumped, while on June 1, from 8 a. m. to 4 p. m., the wind velocity varied from 11 to 15 miles an hour, and not a tankful was pumped. The report gives the number of tankfuls of 141.2 cubic feet which were lifted 12.85 feet each hour during the year, and some interesting conclusions drawn from these records.

Professor King has also made some tests of this mill with a Prony friction brake. The results of these tests, and the indicated horsepowers computed from them, are as follows:

---

[1] Bull. No. 68, Wisconsin Agricultural Experiment Station.

Results of tests of 16-foot geared Aermotor.

Wind velocity per hour.	Direction of wind.	Indicated horse-power.	Average wind velocity.	Average horse-power.	Barometer.	Temperature.
Miles.			*Miles.*		*Inches.*	*Degrees.*
8.4	SW.	0.2715	8.4	0.2715	29.40	0.0
12.0	....	0.5791	} 12.33	0.5858	29.36	0.5
12.4	....	0.7230			29.36	0.5
12.6	....	0.4553			29.36	0.25
13.2	SW.	0.6213	13.2	0.6213	29.40	0.5
14.6	SW.	0.7343	} 14.68	0.8602	29.36	−0.5
14.6	....	0.9449			29.40	0.5
14.8	SW.	0.9016			29.40	1.0
18.6	NW.	2.054	} 18.70	1.873	28.86	9.0
18.8	SW.	1.692			29.40	1.0
21.2	SW.	2.593			29.40	1.5
21.6	W.	3.715	21.55	3.033	29.06	−2.0
21.6	N.	3.227			28.86	7.0
21.8	NW.	2.597			....	....
22.0	W.	4.326	} 22.06	3.652	29.09	1.0
22.0	W.	4.236			29.04	−2.0
22.2	NW.	2.394			....	....
23.0	W.	3.842	} 23.00	3.996	29.07	−1.0
23.0	W.	4.151			29.05	−1.0
24.0	....	5.983	} 24.30	4.768	28.75	6.8
24.6	W.	3.554			29.09	1.5
25.2	W.	4.882	25.2	4.882	29.09	1.5
27.3	N.	4.092	} 27.16	4.471	28.66	6.3
27.0	W.	4.850			29.10	1.5
39.0	N.	5.953	39.0	5.953	28.69	6.5
40.0	....	5.971	40.0	5.971	28.69	6.3

This mill is similar to our 16-foot Aermotor No. 44, described in Part II. It will be seen later that the power found by Professor King is much greater than we have found it for high wind velocities. The probable reason for this difference will be discussed later. It may be stated here, however, that Professor King measured the wind velocity with an anemometer in a fixed position 40 feet due east of the windmill, and it will be seen that the wind was from the west, northwest, or southwest nearly all of the time when the brake tests were being made. (The wind came from the north around by the east to the south only three times while the brake tests were in progress.) The revolving wheel must have interfered with the running of the anemometer and caused it to show a less wind velocity than really existed.

Professor King also determined the work done by the mill in grinding corn. The power of the mill in a given wind velocity can not, however, be judged from these tests, since the grinder load was probably not suited to all wind velocities.

Results of brake tests of 16-foot geared Aermotor.

Wind velocity per hour.	Indicated horse-power.	Wind velocity per hour.	Indicated horse-power.
8 miles	0.25	26 miles	4.82
10 miles	0.40	28 miles	5.14
12 miles	0.56	30 miles	5.40
14 miles	0.78	32 miles	5.61
16 miles	1.08	34 miles	5.76
18 miles	1.62	36 miles	5.87
20 miles	2.39	38 miles	5.95
22 miles	3.31	40 miles	5.97
24 miles	4.31		

Perry's experiments.—From June, 1882, to September, 1883, Mr. T. O. Perry made experiments with 61 windmills, each 5 feet in diameter. The results were published in 1889.[1] Mr. Perry's methods were similar to those employed by Smeaton—he used small wheels moved against still air in a circle of 14 feet radius. His experiments were made on a much larger scale than Smeaton's, however, and his apparatus was more perfect. Smeaton used wheels of European type; Mr. Perry used those of American type. Pl. II is an elevation of Mr. Perry's apparatus, showing the wheel as it revolved about the vertical shaft, driven by an 80-horsepower engine. The power was measured by means of a Prony friction brake placed on a brass cylinder on the wind-wheel shaft. In order to eliminate the effect due to differences in the condition of the air, and get results comparable with one another, Mr. Perry used one of his wheels as a standard with which to compare the others. After the best load for a wheel had been obtained, comparative tests were made with this one and with the standard wheel, by trying first one wheel and then the other until several measurements of each had been taken. The final result of each wheel was the average of eight or ten measurements.

In comparing Smeaton's results with his own, Mr. Perry writes:

We were not able to obtain the best results with weather angles as small as Smeaton's in any of our wheels. Nor did our sail speeds, as compared with wind velocity, nearly approach the speeds obtained by Smeaton. Even our unloaded wheels did not show the sail speed attained by the best of Smeaton's when loaded for maximum work. * * * Our loads at the maximum of work were smaler as compared with the greatest loads, and the speed of revolutions at maximum work as compared with the speeds of unloaded wheels, were smaller for our mills than for Smeaton's.

He states, however, that the general conclusions drawn by Smeaton (see pages 15 to 16) were substantially confirmed by his experiments. The difference between his results and Smeaton's he attributes to the differences in the mills used.

---

[1] Water-Supply and Irrigation Paper U. S. Geol. Survey No. 20.

Some of Mr. Perry's conclusions are as follows:

(1) There is nothing gained by having the sail area more than seven-eighths of the wind area, and there is little gained by having it more than three-fourths of the latter area.

(2) That the power varies as the cube of the wind velocity.

(3) That the load for maximum power varies as the square of the wind velocity.

(4) That the speed of the unloaded wheel increases somewhat faster than the wind velocity.

(5) That the best speed for most of his wheels was about 0.55 per cent of the unloaded speed.

(6) That the conical deflector at the center of the wheel does not increase the power.

(7) That obstructions on the back of sails greatly reduce the power of the mill.

(8) That the speed of wheel No. 48 was increased 48 per cent by removing the strip from the back of each sail.

(9) That a deflector in front of the wheel increased the speed of a slow-moving wheel.

(10) That a deflector in front of the wheel did not increase the speed of a rapidly moving wheel.

(11) That a mast offers more obstruction in front of a wheel than behind it.

EXPERIMENTS BY WRITER.

The tests of windwills described in the following pages were begun by the writer in the summer of 1895. They were continued during the summer of 1896, with much better facilities than during the previous season. The results obtained to that time were published in Water-Supply and Irrigation Paper No. 8, entitled Windmills for Irrigation. Since then the work has been continued as time could be spared—mainly during a portion of three summer vacations. The work of the summer of 1896 was confined mainly to pumping mills. The tests show what each windmill and its pump were actually doing under certain conditions of load, lift, etc. They do not show what the mill might do under other conditions. It was evident that the useful work of a mill varied with its load and the efficiency of the pump. The latter could not well be ascertained. It was, therefore, thought best to confine the tests principally to power mills, in which the unknown factor of pump efficiency is not present, and where the load on the mill can easily be varied. This has enlarged the scope of the work, making it cover windmills for power as well as those for irrigation.

Many of our tests of pumping mills were made in the vicinity of Garden, Kansas. Perhaps nowhere in the United States is irrigation from wells by the use of windmills carried to the same extent as there, where may be found hundreds of windmill pumping plants furnishing water to irrigate from 1 to 15 acres each and lifting from 3 to 14.5 quarts per stroke to a height of from 10 to 45 feet, as well as large steel mills running day and night, when the wind is strong enough, and working pumps of the best kind.

SCOPE OF TESTS.

There are many makers of American windmills, and with the great variety of mills in use—no two are alike, though in some cases the difference is slight—it was impossible to test a mill of each type. It was our purpose to test only the mills that were in good working order and subject to good wind exposure, and which would add new data to that already obtained or confirm in some particular that previously secured. In some cases two or more mills of the same size and make were tested to show as far as possible the effect of pump, well, etc. on the useful work. The parts of mill and pump on which the power depends were carefully measured. The temperature and barometric pressure were observed in each case and the mean given. The discharge of pump per stroke was measured when possible, and the diameter of cylinder and length of stroke are given, so that the discharge can be compared with the figured displacement. The lift was measured whenever the surface of the water in the well could be reached with a tapeline. The number of strokes of the pump per mile of wind was found for velocities from 6 or 8 miles to 20 or 30 miles an hour. In some cases the number of strokes is given when no water was being pumped. In fact, there was collected for each mill as much data as it was conveniently possible to obtain which would in any way affect the power of the mill or be of interest.

PUMPING MILLS.

The essential difference between pumping and power mills is that in the former there is a pump rod with an up and down motion, while in the latter there is a vertical rotating shaft. The former is usually geared back 2 or 3 to 1, while the latter is generally geared forward 6 or 8 to 1. The ordinary pumping mill, such as is used for stock purposes, is lighter than the power mill, but the irrigating mill is of nearly the same weight as the power mill. The larger power mills have a pumping attachment, so as to work a pump as well as a grinder or other machine.

WELLS NEAR GARDEN, KANSAS.

A brief description of the water supply and wells of this locality may be helpful in considering what follows.

The water is found in sand and gravel at distances below the surface varying from 8 to 40 feet. This material is in layers of variable thickness and different degrees of coarseness, ranging from fine sand to large gravel. It is overlain by a layer of sandy clay, which in some places will for years stand vertical without any support; in other places there is very little clay in this layer. The wells are usually 3 to 4 feet square, and are cased with wood through the top sandy clay to the water-bearing sand; then a wood or galvanized-iron casing from 12 inches to 3 feet in diameter extends down from 8 to 20 feet into the sand to a layer of gravel. Where this latter casing is large, three or more galvanized-iron pipes 6 to 12 inches in diameter are put down in the bottom of it, and these sometimes have wire gauze over their tops to keep down the sand; they also have perforations about one-fourth of an inch in diameter for a distance of 2 feet or more from the bottom to admit the water. In many cases, instead of this small open well, the supply pipe is on a well point having the same diameter as the supply pipe, its length varying with the diameter. These well points have not given satisfaction and are being replaced by open wells.

On examining well points that have been used for a time it was found that many of the little openings through which water is admitted to the pump had become filled with fine grains of sand, thus reducing the area. Although this water area was of the proper amount when the well was new, it becomes too small after the well has been used for a time or after it has stood without being used. If this area is too small to allow the free passage of water into the pump, an added load is put on the latter.

PUMPS.

Nearly all of the pumps in use in the vicinity of Garden are of the reciprocating-piston type. Fig. 8 shows the Stone pump, manufactured by R. G. Stone, of Garden. This pump is made in three sizes— 6 inch, 8 inch, and 10 inch, these dimensions being the approximate diameter of the discharge pipe. The diameter of the cylinder is less than the diameter of the pipe by twice the thickness of the brass lining. The valves (shown in fig. 9) are of the latest pattern. The plunger valve is of the single-flap or clack variety and the check valve is of the disc variety, made so that the water can pass up the center as well as around the sides. In an earlier form of this pump the plunger valve is of the double-flap or butterfly type and the check valve of the lift type, but with no opening at the center. Probably nine-tenths of the pumps in use near Garden are of the Stone variety.

Fig. 3 shows the Gause pump, one of the first pumps used there for irrigating purposes. It is more expensive than the Stone pump, and is not now so much used. Fig. 16 shows the cylinder of an 8-inch Frizell pump, a few of which are in use. Fig. 5 is a sectional view of the Woodmanse pump, which is used with mill No. 2. Pl. X, B, shows a crude homemade pump called the "water elevator." One of these is in use in Garden.

The efficiency of reciprocating pumps like those described varies directly with the lift, inversely with the number of strokes per minute, and with the design of the pump. For lifts of 10 or more feet and not more than 30 strokes per minute the efficiency in a good pump should be at least 70 per cent.

Prof. O. P. Hood [1] has measured the efficiency of two Frizell pumps— one 6 inch, with 14.1 inches stroke, like that shown in fig. 16, and one 4 inch, with 24 inches stroke, having a butterfly discharge valve. He found that for a 7.7-foot lift the efficiency of the 6-inch pump dropped from 75 per cent to 63 per cent as the number of strokes increased from 10 to 60 per minute; that for a 22.7-foot lift it decreased from 86 per cent to 82 per cent for the same range of speed; and that for a 37.8-foot lift it decreased from 84 per cent to 82 per cent, while the number of strokes increased from 10 to 50 per minute. The efficiency of the 4-inch pump dropped from 66 per cent to 60 per cent for a 12.8-foot lift, and from 83 per cent to 73 per cent for a 37.6-foot lift, as the number of strokes increased from 10 to 50 per minute.

The valve area should be not less than 30 per cent of the cylinder area. The cylinder should be placed as near the water as possible; if it is more than 25 feet above the water, and the number of strokes is 30 or more, the cylinder will not fill properly, and pounding will result.

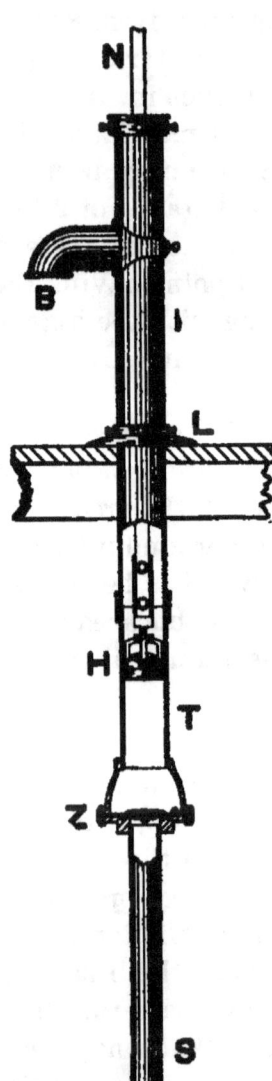

FIG. 3.—Gause pump. *N*, plunger; *B*, spout; *F*, discharge pipe; *H*, plunger; *T*, cylinder; *Z*, enlarged valve opening and check valve; *S*, suction pipe.

INSTRUMENTS AND METHODS.

The wind velocity was measured with a United States Weather Bureau cup anemometer, each mile of wind being recorded electrically by one pen of a 2-pen register. By means of a little device fastened to the pump an electric circuit is closed at each stroke of the pump and a record made by a recorder. Another electric circuit, leading from the recorder to the other pen of the register, is closed at each hundred strokes of the pump and a record made on the register. Hence the graphic record of the register shows the number of miles of wind in any given time, also the number of hundred strokes of the pump in the same time. The anemometer was held on a pole at the height of the axis of the wheel of the windmill. The pole was made so that its length could be increased at will from 25 to 50 feet. The anemometer on the pole is shown in Pls. III and XI.

---

[1] Water-Supply and Irrigation Paper U. S. Geol. Survey No. 14.

The discharge of the pump per stroke was ascertained by catching the water for several strokes in a tub and measuring it with a quart measure. In a few cases it was found to vary with the number of strokes per minute. Where it varied the discharge given is for a nearly maximum speed of pump.

The lift, or distance from the surface of the water in the well to the center of the water column as it leaves the discharge pipe, was measured when the pump was working quite rapidly. For pumps on well points it was estimated from the depth to water when the point was put down, making an allowance for the lowering of the water.

Each mill tested is described and the results of the tests given in tabular form. Nearly all of the mills are illustrated.

The number of strokes of the pumps per mile of wind and the horsepowers of the mills are in most cases explained by diagrams, which show at a glance the facts which otherwise can be comprehended only by a careful analysis of the tables. In these diagrams (figs. 6, 10, 11, 12, 13, 15, 17, 18, 19, and 21) the relation between the wind movement, in miles per hour, and the number of strokes made by the pump while the wind was moving over 1 mile is shown by the curved line. The space from left to right is proportional to the number of strokes of the pump. The data expressed by these diagrams were obtained directly from the record given by the anemometer register.

In explanation we will assume that the pen connected with the anemometer makes three short marks (3 miles) in fifteen minutes, indicating a mile in five minutes, or at the rate of 12 miles an hour. At the same time the other pen connected with the pump, and registering each 100 strokes, makes, say, two short marks, showing that the pump has made 200 strokes for this 3 miles of wind movement, or 67 strokes to the mile. This fact is entered on the diagram by a small circle placed at a distance from the right which corresponds to a wind velocity of 12 miles an hour, and at a distance from the bottom which corresponds to 67 strokes of the pump. In this way each observation is indicated. When the points have been plotted, the smooth curve is sketched so as to occupy an intermediate position among them.

In order to obtain the number of strokes more accurately than by measurement on the register sheet, they were actually counted for a considerable number of observations in each test. The number of strokes per minute is obtained by dividing the number of strokes per mile of wind by the number of minutes required to make the mile. For example: If the number of strokes per mile in a 12-mile wind (which requires 60 ÷ 12, or five minutes to make a mile) is 90, then 90 ÷ 5 = 18, the number of strokes per minute. The number of gallons raised per minute is found by multiplying the number of gallons per stroke by the number of strokes per minute. The horsepower of the mill in any wind velocity is found by multiplying the number of

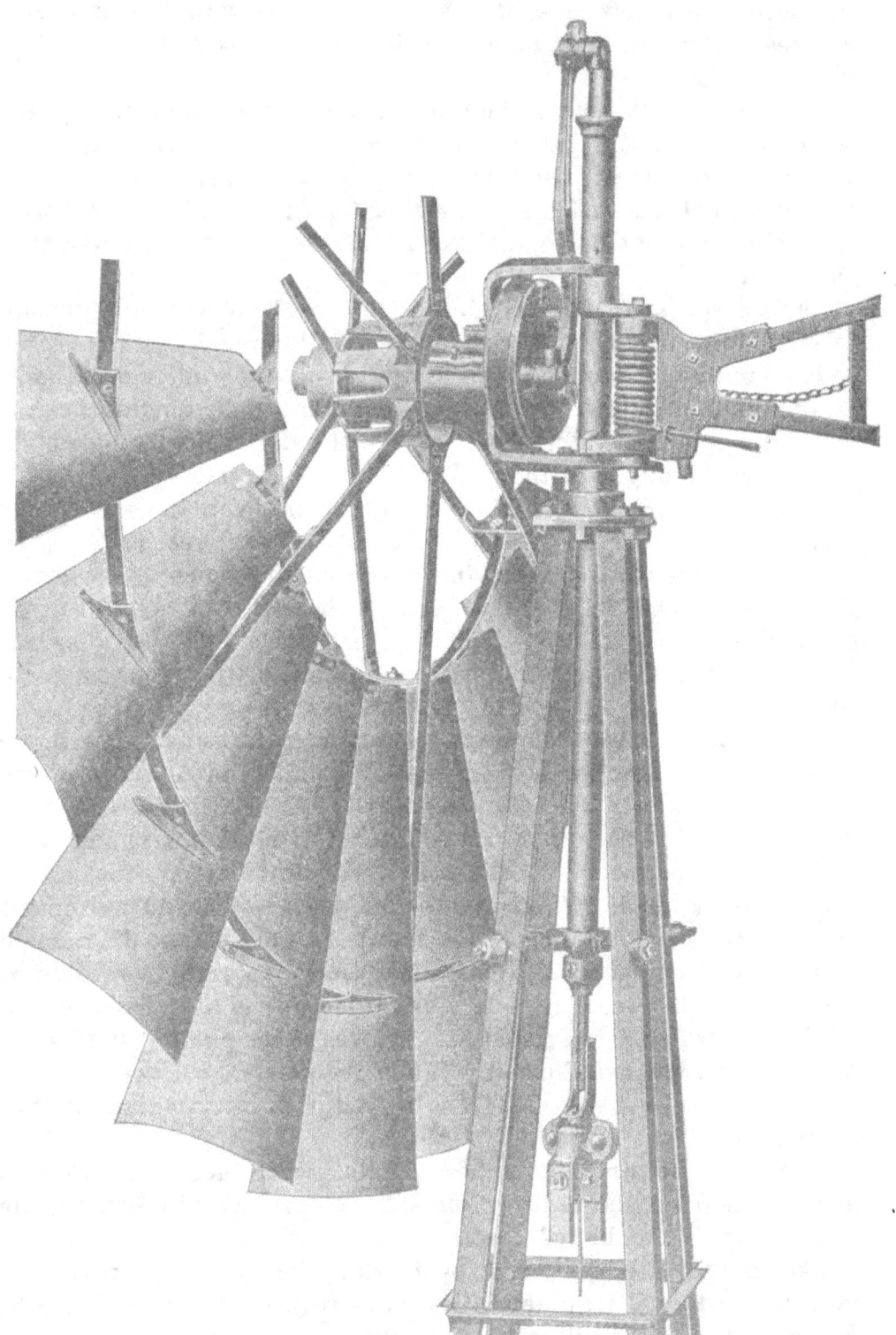

FIG. 4.—Working parts of Woodmanse Mogul.

gallons per stroke by the number of strokes per minute, then by 8.3 pounds (the weight of one gallon of water), then by the lift, in feet, and dividing the product by 33,000 (the number of foot-pounds in a horsepower), or by the following formula:

Horsepower = $nqgh \div 33{,}000$, where n = number of strokes per minute, g = the weight of one gallon of water, q = the number of gallons per stroke of pump, h = the lift, in feet. The pump load is the weight of water lifted per stroke multiplied by the lift, or height to which it is raised. The number of revolutions of the wind wheel per minute is found by multiplying the number of strokes per minute by the number of revolutions per stroke.

PUMPING MILLS TESTED.

Mill No. 1.—The tests of this mill were preliminary or experimental, being made for the purpose of perfecting the instruments employed, and were not completed for discussion.

Mill No. 2.—This is a 12-foot Woodmanse Mogul, manufactured by the Woodmanse-Hewitt Manufacturing Company, of Freeport, Illinois. Pl. III shows the mill, tower, pump, and pond, and fig. 4 the working parts. The tower is of steel, 50 feet high to the axis of the wheel. The wind exposure on the north, is not good, the mill being 115 feet south of a large barn. The wheel has 30 curved sails, each 36 by 13 by 5.5 inches,[1] set at an angle of 30° (angle of weather) with the plane of the wheel. It is

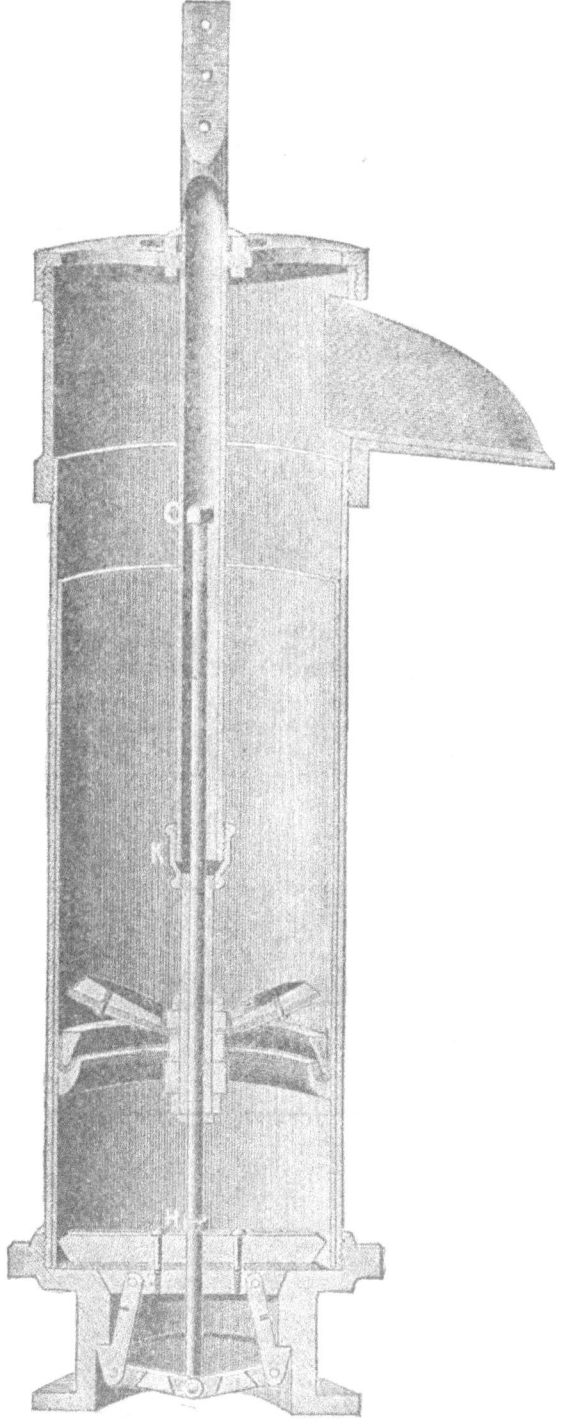

Fig. 5.—Sectional view of Woodmanse pump.

back-geared, 3 to 1, and held in the wind by a spring. The pump

---

[1] In this expression 36 is the length of the sail, 13 the width of sail at the outer end, and 5.5 the width of sail at the inner end.

also is of Woodmanse make, and is shown, in section, in fig. 5.   The cylinder is 9.5 inches in diameter, the supply pipe 5.625 inches in diameter, the length of stroke 12 inches.   The well is 3¾ feet by 3¾ feet to the water, a distance of 14 feet.   At that point a 12-inch galvanized-iron pipe is put down 20 feet, forming a small open well. The lift at the time of test was 17¾ feet and the discharge per stroke 14½ quarts.   The mean barometric pressure was 26.98 inches, and the mean temperature 94° F.   The cost of mill, tower, pump, and well was about $210.   The results of the tests are as follows:

Results of tests of mill No. 2—12-foot Woodmanse Mogul.

[Load per stroke, 536.2 foot-pounds.]

Wind velocity per hour.	Revolutions of wind wheel per minute.	Strokes of pump per minute.	Gallons pumped per minute.	Useful horsepower.
12 miles	15.6	5.2	18.8	0.085
16 miles	48.0	16.0	58.0	0.260
20 miles	60.9	20.3	73.6	0.322
25 miles	69.9	23.3	84.5	0.379
30 miles	75.9	25.3	91.7	0.411

The curve shown in fig. 6 is for a moderately loaded 12-foot mill (536.2 foot-pounds per stroke).   It starts at a wind velocity of 11

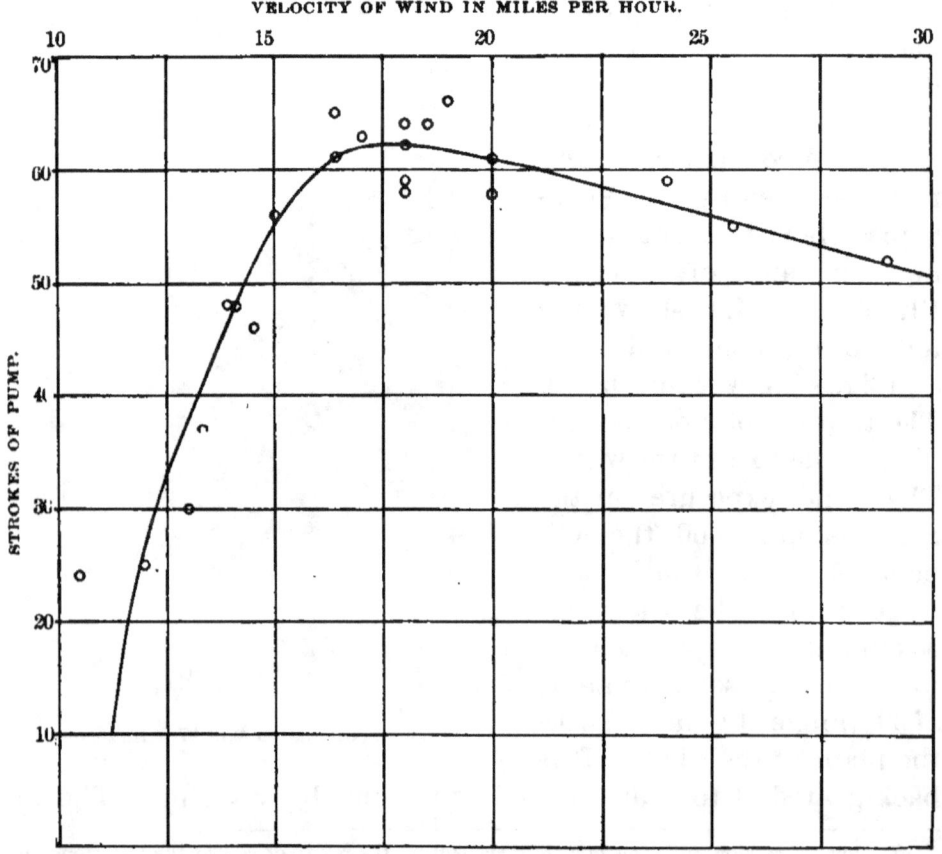

VELOCITY OF WIND IN MILES PER HOUR.

FIG. 6.—Diagram showing results with mill No. 2—12-foot Woodmanse Mogul.

miles an hour. It ascends very rapidly, reaching a maximum at 18 miles an hour, and giving 60 strokes to the mile. The rest of the curve to 30 miles has a gentle slope. The number of strokes per minute increases from about 5 at 12 miles to about 25 at 30 miles an hour, and will continue to increase to probably 28 a minute in a 40-mile wind.

Mill No. 3.—This is a 12-foot Aermotor manufactured by the Aermotor Company, of Chicago, Illinois. Pl. IV shows the mill with its tower, pump, and pond, and fig. 7 shows its working parts. This mill had been in use about one year at the time of test, and all of the parts were in good working order. The tower is of wood, the axis of the wheel being 30 feet above the ground. The exposure is very good. The wheel has 18 curved sails, each 44 by 18¾ by 7¾ inches, set at an angle of 31° to the plane of the wheel. It is back-geared, 3½ to 1, and is held in the wind by a spring. The pump is of the Stone type, shown in figs. 8 and 9; the check valve is of the single-flap variety, the plunger valve of the double-flap variety. The cylinder is 9½ inches in

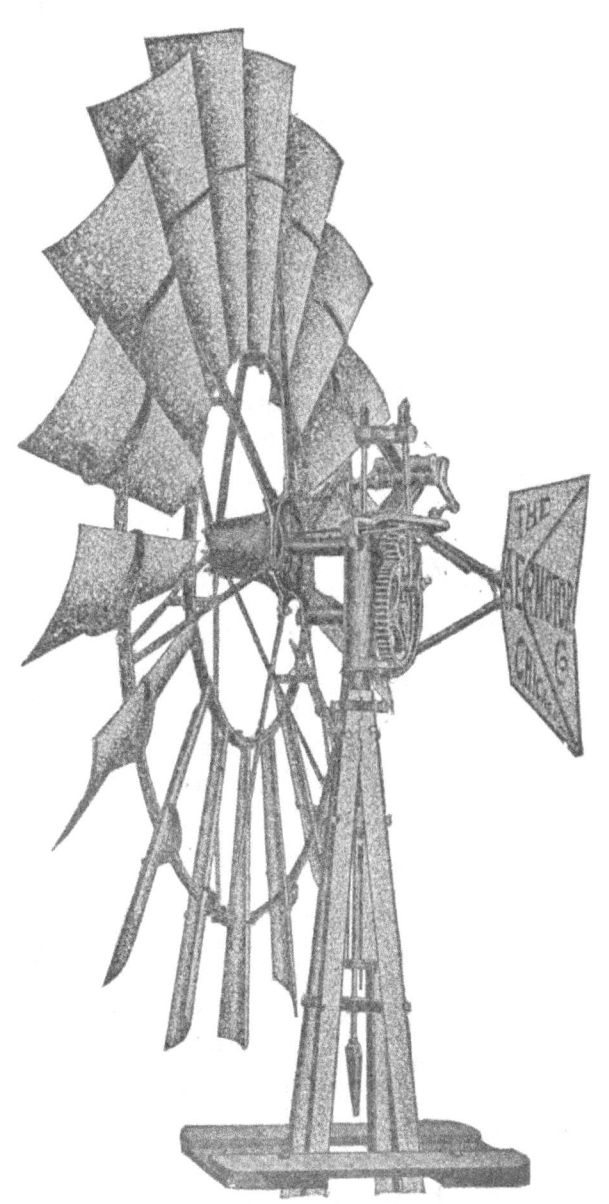

FIG. 7.—Working parts of Aermotor

diameter, the supply pipe 4 inches in diameter, and the discharge pipe 10 inches in outside diameter; the length of stroke is 12 inches and the discharge per stroke 14½ quarts. The well is 4 feet by 4 feet to a depth of 8 feet—nearly down to water. From that point to a depth of 18 feet it is 3 feet in diameter; and from there three pipes, 12 inches in diameter, extend down 5 feet farther. The lift at the time of test was 13¾ feet, the barometric pressure 27.2 inches, and the temperature 85° F. The water is pumped into a pond

80 feet by 75 feet, and a depth of 22 inches can be drawn off. The cost of the plant, including mill, tower, pump, and pond, was $145. The greatest number of strokes per minute is probably 27 or 28 in a 40-mile wind. In general appearance this curve is seen to resemble that of mill No. 2, but it is about 4 miles farther to the left, due to lighter load. The results of the tests are as follows:

Results of tests of mill No. 3—12-foot Aermotor.

[Load per stroke, 415.3 foot-pounds.]

Wind velocity per hour.	Revolutions of wind wheel per minute.	Strokes of pump per minute.	Gallons pumped per minute.	Useful horse power.
8 miles	17.7	5.3	19.3	0.067
12 miles	40.0	12.0	43.5	0.151
16 miles	54.7	16.4	59.5	0.207
20 miles	66.0	19.8	71.8	0.250
25 miles	77.0	23.1	83.6	0.291
30 miles	83.3	25.0	90.6	0.315

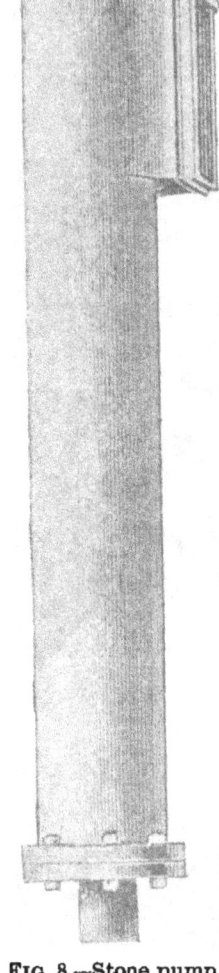

FIG. 8.—Stone pump.

The curve shown in fig. 10 is for a rather lightly loaded mill (415.3 foot-pounds per stroke). It starts at a velocity of 6 to 7 miles an hour, ascends less rapidly than the one shown in fig. 6, attains a maximum at about 15 miles an hour, when the number of strokes per mile is 62, and then descends slowly, reaching 50 strokes at 30 miles. The number of strokes per minute increases from about 5 at 8 miles to 25 at 30 miles.

Mill No. 4.—This mill, shown in the foreground of Pl. V, is an 8-foot Ideal windmill manufactured by the Stover Manufacturing Company, of Freeport, Illinois. It had been in use about one year, and all of the parts were in good condition. The tower is of wood, the axis of the wheel being 48 feet above the ground. The wheel has 15 sails, each 16¼ by 7 by 30 inches, set at an angle of 29° with the plane of the wheel. It is back-geared, 2½ to 1, and is held in the wind by a spring. The pump is of the Stone make. The diameter of the discharge pipe is 5⅜ inches, of the supply pipe 3 inches. The length of stroke is 8 inches. The plunger and check valves are of the single-flap variety. The well is 2¾ feet by 2¾ feet down nearly to water—a depth of 5¼ feet. The 3-inch supply pipe extends down to a depth of 14 feet, and on the end of it is a 3-inch well point 6 feet long. The lift may vary from 8½ to 20 feet. It was probably about 12 feet at the time of tests. The discharge per stroke was 2 quarts. The mean barometric pressure was 27.19 inches, the mean temperature 83° F.　The water is pumped into a pond 115 feet by 31 feet and 3 feet deep.　The cost of the plant, including mill, pump, well, and pond, was $80.　The results of the tests are as follows:

Results of tests of mill No. 4—8-foot Ideal.

[Load per stroke, 50 foot-pounds.]

Wind velocity per hour.	Revolutions of wind wheel per minute.	Strokes of pump per minute.	Gallons pumped per minute.	Useful horse-power.
12 miles	25.5	10.2	5.1	0.015
16 miles	48.2	19.3	9.6	0.029
20 miles	63.2	25.3	12.6	0.038
25 miles	70.3	28.1	14.1	0.043
30 miles	62.5	25.0	12.5	0.038

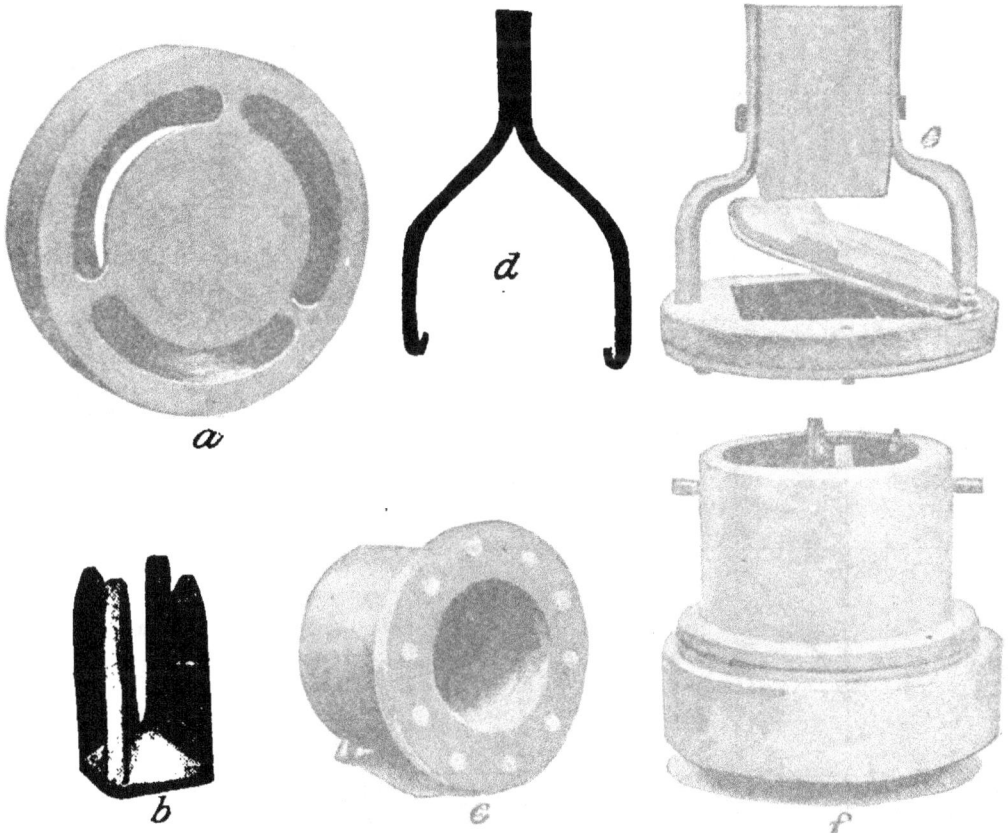

FIG. 9.—Details of Stone pump: *a*, Lower valve seat; *b*, ring guide to lower valve; *c*, lower or check valve; *d*, hook for removing lower valve; *e*, plunger and valve; *f* is *a*, *b*, and *c* combined.

The curve shown in fig. 11, although for a rather lightly loaded mill—50 foot-pounds per stroke—shows that the mill starts in a 10-mile to an 11-mile wind.　The maximum is reached at 19 miles, with a speed of 78 strokes.　The right side of the curve is quite steep, a characteristic of this make of mill.　Mill No. 18 is the same size and make as this mill, and yet with a load of 89.2 foot-pounds it starts in a 7-mile to an 8-mile wind, reaching a maximum at about 13 miles, at

VELOCITY OF WIND IN MILES PER HOUR.

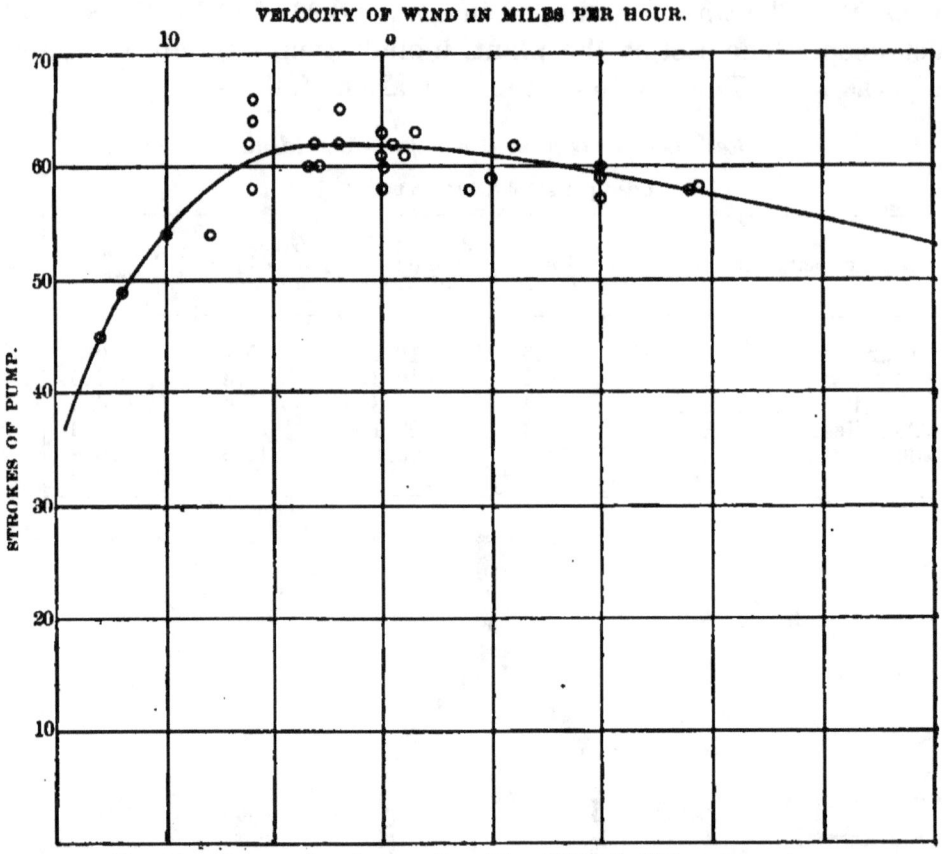

FIG. 10.—Diagram showing results with mill No. 3—12-foot Aermotor.

VELOCITY OF WIND IN MILES PER HOUR.

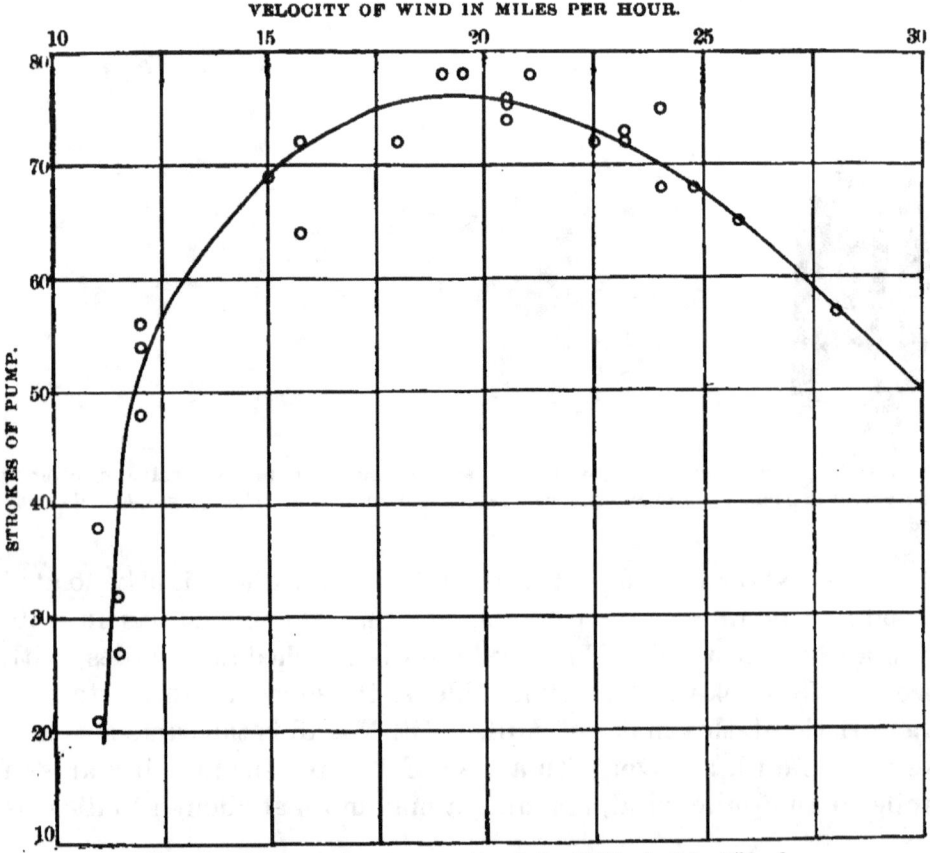

FIG. 11.—Diagram showing results with mill No. 4—8-foot Ideal.

a speed of 104 strokes per mile. A second test, when the spring that holds the wind wheel in the wind was tightened somewhat, gave the maximum at a velocity of about 15 miles, with a pump speed of about 114 strokes. The difference appeared to be due to the difference in pumps and wells. The rapid fall in the curve to the right of the highest point is due to the easy governing of the mill.

Mill No. 5.—This mill, shown in the background of Pl. V, is an 8-foot Aermotor manufactured by the Aermotor Company, of Chicago, Illinois. The tower is of wood, and is 28.5 feet high to the axis of wheel. The exposure is good, and all of the parts were in good working order at the time of tests, the plant having been in use about one year. The wheel has 18 curved sails, each 30 by 12½ by 5½ inches, making an angle of 29½° with the plane of the wheel. It is back-geared, 3⅓ to 1. The pump is of the Stone make. The discharge pipe is 6 inches in diameter, the supply pipe 3 inches in diameter. The valves (check and plunger) are of the single-flap variety. The length of stroke is 8 inches. The well is 4 feet by 4 feet to water, a depth of 10.5 feet. A 12-inch wooden curb extends 12 feet farther into the sand and gravel. The discharge per stroke was 3½ quarts, and the lift 13 feet. The cost of plant, including pond, was $80.

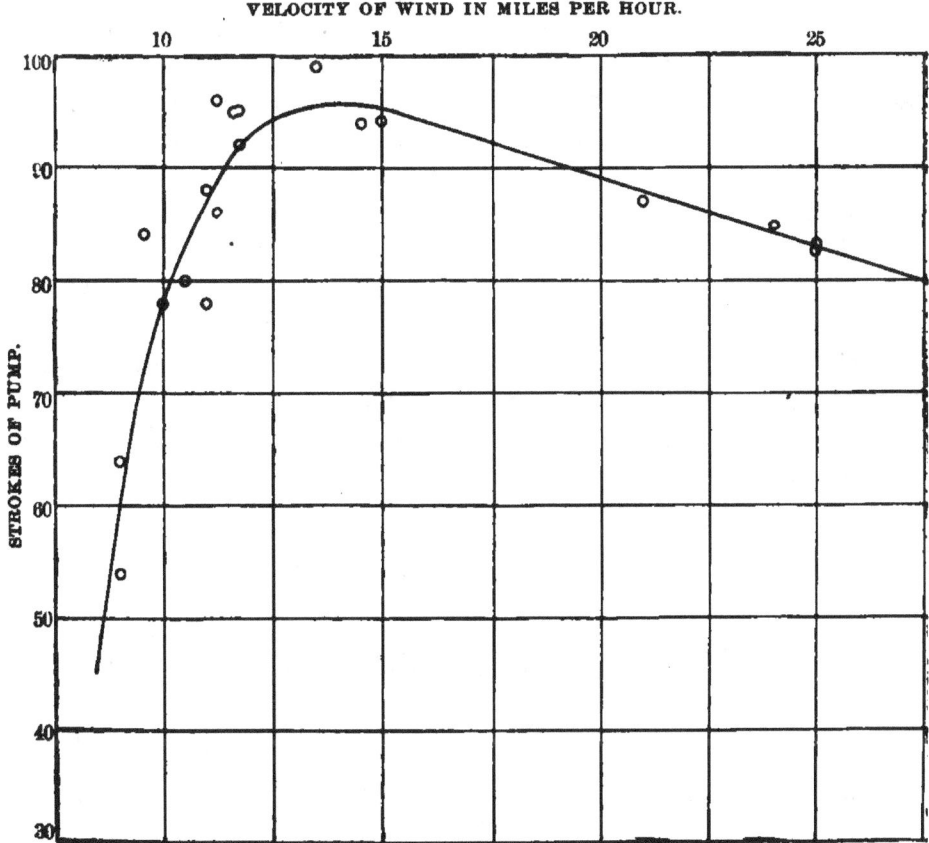

Fig. 12.—Diagram showing results with mill No. 5—8-foot Aermotor.

The results of the tests of mill No. 5 are as follows:

Results of tests of mill No. 5—8-foot Aermotor.

[Load per stroke, 94.9 foot-pounds.]

Wind velocity per hour.	Revolutions of wind wheel per minute.	Strokes of pump per minute.	Gallons pumped per minute.	Useful horse-power.
12 miles..............	62.0	18.6	16.3	0.053
16 miles..............	83.7	25.1	22.0	0.072
20 miles......	99.3	29.8	26.1	0.086
25 miles.............	115.3	34.6	30.3	0.099
30 miles	128.3	38.5	33.7	0.111

This 8-foot mill, with a load of 95 foot-pounds, is seen to start in an 8-mile to a 9-mile wind (fig. 12), reaching a maximum at 13 to 15 miles, with 95 strokes per mile. At 30 miles an hour it is making 77 strokes per mile, or 39 strokes per minute. This curve indicates a rather heavily loaded mill.

Mill No. 6.—This mill, shown in the background of Pl. IV, is an 8-foot Gem, manufactured by the United States Wind Engine and Pump Company, of Kansas City, Missouri. The working parts of the mill are shown in Pl. VI. The exposure was good and all of the parts were in good working order, the mill having been in use only about one year at the time of tests. The wheel has 24 curved sails, each $30\frac{1}{2}$ by 10 by $4\frac{1}{2}$ inches, set at an angle of 35° with the plane of the wheel. It is back-geared, 3 to 1. The wheel is held in the wind by means of a weight. The pump is of the Stone make. The discharge pipe is 6 inches in diameter, the supply pipe 4 inches in diameter. The length of stroke is 8 inches. The well is open to the water—a depth of $6\frac{1}{2}$ feet. The supply pipe is on a well point, the end of which is 16 feet below the surface of the ground. The lift was $9\frac{7}{12}$ feet and the discharge per stroke 3.9 quarts. The tower is of wood, and is 24 feet high to the axis of the wheel. The mean barometric pressure was 27.02 inches, the mean temperature 85° F. The plunger valve is of the double-flap variety, and the check valve of the single-flap variety. The results of the tests are as follows:

Results of tests of mill No. 6—8-foot Gem.

[Load per stroke, 77.6 foot-pounds.]

Wind velocity per hour.	Revolutions of wind wheel per minute.	Strokes of pump per minute.	Gallons pumped per minute.	Useful horse-power.
12 miles.............	37.2	12.4	12.1	0.029
16 miles.............	53.7	17.9	17.5	0.042
20 miles.............	66.0	22.0	21.5	0.051
25 miles.............	76.2	25.4	24.7	0.059
30 miles.............	85.5	28.5	27.8	0.065

Comparing the number of strokes per minute of mills Nos. 4, 5, and 6, it is seen that although No. 5 is carrying a much heavier load than either of the other mills, it makes more strokes and does much more work at all velocities.

Mill No. 7.—This is a 12-foot Aermotor similar to mill No. 3. The tower is of steel, having a height of 31 feet to the axis of the wheel. The exposure was good and all of the parts were in good working order, the plant having been in use less than one year when tests were made. The pump is of the Stone make and is like that of mill No. 3, except that the check valve is of the solid-lift variety. The lift was 15½ feet and the discharge 14.3 quarts per stroke. The water is pumped into a pond 135 feet by 50 feet by 2½ feet.

Comparing the results of the tests of this mill with those of mill No. 3, it is seen that the latter is somewhat more heavily loaded than the former and makes a few less strokes per minute, but that its horsepower is a little greater. The effect of the larger load is shown.

The results of the tests of mill No. 7 are as follows:

Results of tests of mill No. 7—12-foot Aermotor.

[Load per stroke, 461.9 foot-pounds.]

Wind velocity per hour.	Revolutions of wind wheel per minute.	Strokes of pump per minute.	Gallons pumped per minute.	Useful horsepower.
12 miles	38.0	11.4	40.7	0.160
16 miles	52.7	15.8	56.5	0.221
20 miles	63.3	19.0	67.9	0.266
25 miles	73.7	22.1	79.0	0.309
30 miles	78.3	23.5	84.0	0.329

Mill No. 8.—This is a 10-foot Star wooden mill, manufactured by Bradley, Wheeler & Company, of Kansas City, Missouri. The tower is of wood, and the axis of the wheel 35½ feet above the ground. The water is pumped into an elevated tank 20 feet above the surface of the ground and is used for irrigation. The wheel has 60 plane sails, each 37 by 5 by 2¾ inches, set at an angle of 33° to the plane of the wheel. It is held in the wind by means of a weight. It is not back-geared, a stroke of the pump being made to each revolution of the wheel. The supply pipe is 2 inches in diameter and terminates in a well point, the end of which is 18 feet below the surface of the ground. The discharge pipe is 1¼ inches in diameter. The cylinder is 3 inches in diameter, the length of stroke 5 inches. The lift may vary between 28½ and 37 feet. It was estimated to be about 30 feet at the time of measurement. The discharge per stroke was 0.24 quart. The cylinder leaked some at the time of tests. After a new cylinder was put in the discharge per stroke was increased to 0.40 quart. The mean barometric pressure was 27.04 inches, the mean temperature 78° F. The results of the tests are as follows:

Results of tests of mill No. 8—10-foot Star wooden mill.

[Load per stroke, 15 foot-pounds.]

Wind velocity per hour.	Revolutions of wind wheel per minute.	Strokes of pump per minute.	Gallons pumped per minute.	Useful horse-power.
8 miles	28.0	28.0	1.7	0.013
12 miles	30.0	30.0	1.8	0.014

These results show the effect of the very light load and the readiness with which the wind wheel turns out of the wind. It makes 28 strokes per minute in an 8-mile wind and less than that in a 16-mile or higher wind.

Mill No. 9.—This is a 16-foot Aermotor. The tower is of steel, and the axis of the wheel is 30 feet above the ground. The wheel has 18

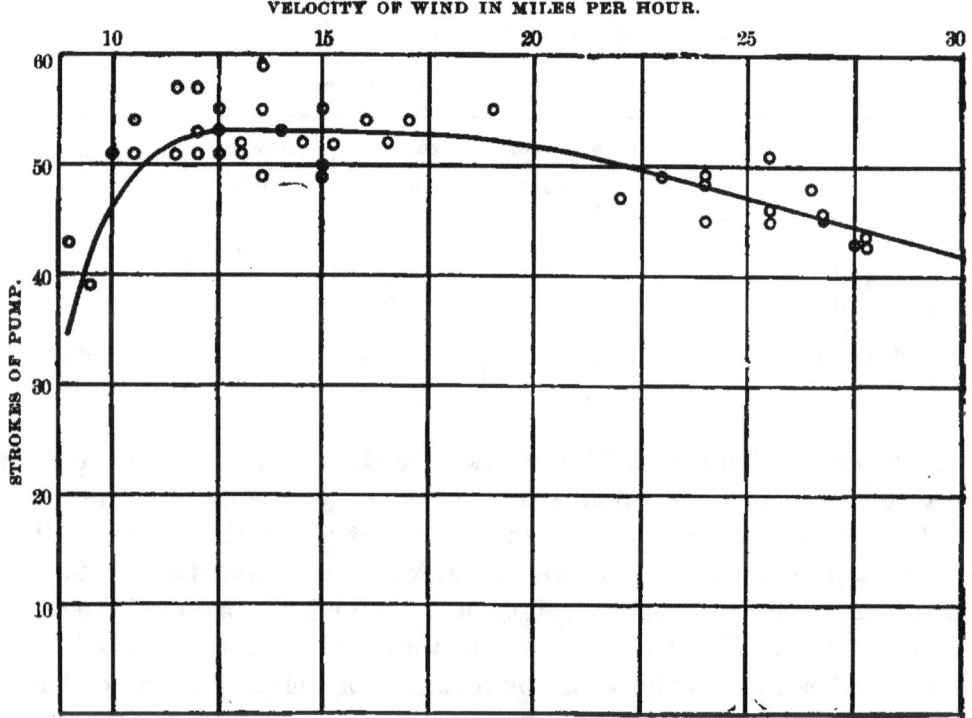

FIG. 13.—Diagram showing results with mill No. 9—16-foot Aermotor.

curved sails, each 59 by 25¾ by 10½ inches, set at an angle of 30° with the plane of the wheel. It is back-geared, 3 to 1. The discharge pipe is 12 inches in diameter, the supply pipe 6 inches in diameter, the cylinder 8 inches in diameter. The stroke is 16 inches. The well is 4 feet by 6 feet to a depth of 23 feet, 2 feet by 2 feet for the next 8 feet, and 18 inches in diameter for the next 14 feet. The water was 39½ feet below the surface of the ground.. The lift was 44¼ feet and the discharge per stroke 11 quarts.   The check valve is of the single-flap variety and the plunger valve of the double-flap variety.   The mean barometric pressure was 27.04 inches, and the mean temperature 93° F.   This plant had been in use about three years.   The results of the tests are as follows:

Results of tests of mill No. 9—16-foot Aermotor.

[Load per stroke, 1,013 foot-pounds.]

Wind velocity per hour.	Revolutions of wind wheel per minute.	Strokes of pump per minute.	Gallons pumped per minute.	Useful horse-power.
12 miles	31.8	10.6	29.1	0.325
16 miles	42.3	14.1	38.8	0.433
20 miles	51.6	17.2	47.3	0.548
25 miles	58.8	19.6	53.9	0.601
30 miles	63.0	21.0	57.7	0.644

The curve shown in fig. 13 starts at a wind velocity of 8 to 9 miles, and reaches a maximum at 13 miles, with a speed of 53 strokes per mile.   From that point to a velocity of about 19 miles the curve is nearly horizontal; after 19 miles it descends slowly to 32 miles, with 38 strokes per mile.   The speed increases from 11 strokes per minute at 12 miles to 21 strokes per minute at 30 miles an hour.

Mill No. 10.—This is an 8-foot Ideal.   The tower is of wood, the axis of the wheel being 30 feet above the ground.   The wheel has 15 curved sails, each 31 by 19 by 7 inches, set at an angle of $29\frac{1}{2}°$ with the plane of the wheel.   It is back-geared, $2\frac{1}{2}$ to 1.   The supply pipe is $1\frac{1}{2}$ inches in diameter, the cylinder $2\frac{1}{2}$ inches in diameter.   The pump is a common hand pump, with lift valve of the flap form and plunger of the lift variety.   The valves leak some, as the discharge is greater when the pump is working rapidly than when it is working slowly.   The supply pipe is on a well point 2 feet long and $1\frac{1}{2}$ inches in diameter, the lower end of which is 50 feet below the surface of the ground.   The lift was 33 feet and the discharge per stroke one-third of a quart when pumping quite rapidly.   The mean barometric pressure was 26.94 inches, the mean temperature 97° F.   The results of the tests are as follows:

Results of tests of mill No. 10—8-foot Ideal.

[Load per stroke, 22.8 foot-pounds.]

Wind velocity per hour.	Revolutions of wind wheel per minute.	Strokes of pump per minute.	Gallons pumped per minute.	Useful horse-power.
8 miles	35.2	14.1	1.2	0.010
12 miles	62.5	25.0	2.1	0.017
16 miles	83.2	33.3	2.8	0.023
20 miles				0.032

Mill No. 11.—This is a 12-foot Ideal, the working parts of which are shown in fig. 14. The tower is of steel, the axis of the wheel being 30 feet above the ground. The exposure was good, and all of the

FIG. 14.—Working parts of mill No. 11—12-foot Ideal.

parts were in good working order when mill was tested. The wheel has 21 curved sails, each 31 by 19 by 7 inches, set at an angle of $29\frac{1}{2}°$ to the plane of the wheel. It is back-geared, $2\frac{1}{4}$ to 1, and the wheel is held in the wind by a spring. The discharge pipe is 8 inches in diameter, and the length of stroke is 12 inches. The supply pipe consists of two 3-inch pipes 14 feet long, each terminating in a 3-inch well point 3 feet long. The valves (check and plunger) are of the single-flap variety. The water was 39 feet below the surface of the ground. The lift, as nearly as could be ascertained at the time of measurement, was 45 feet, the discharge per stroke 9 quarts. The water is pumped into a pond 60 feet by 40 feet by 6 feet. This plant had been in use about three years. The mean barometric pressure was 26.91 inches, the mean temperature 91° F. The results of the tests are as follows:

Results of tests of mill No. 11—12-foot Ideal.

[Load per stroke, 843.7 foot-pounds.]

Wind velocity per hour.	Revolutions of wind wheel per minute.	Strokes of pump per minute.	Gallons pumped per minute.	Useful horse-power.
12 miles	12.0	4.8	10.8	0.123
16 miles	31.7	12.7	28.6	0.325
20 miles	47.0	18.8	42.3	0.481
25 miles	58.2	23.3	52.5	0.600
30 miles	62.5	25.0	56.2	0.639

Mills Nos. 9 and 11 pump water into the same pond from the same depth. It will be seen from these results that for wind velocities of

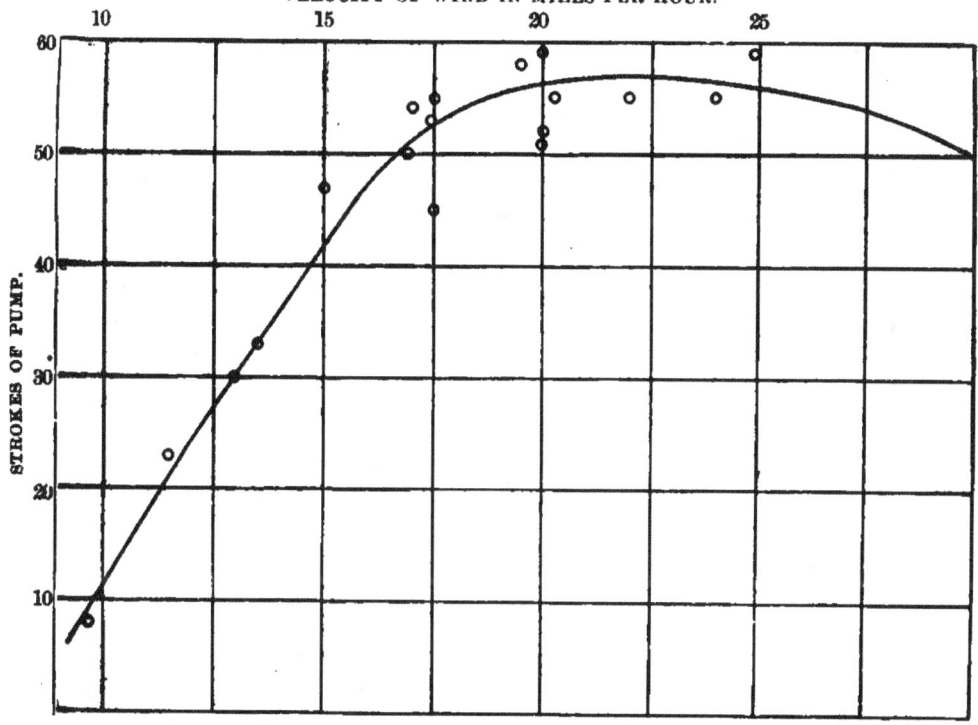

FIG. 15.—Diagram showing results with mill No. 11—12-foot Ideal.

20 miles or more the 12-foot mill is pumping nearly as much water as the 16-foot mill; for velocities of 12 miles or less the 16-foot mill is pumping much more water than the 12-foot mill.

This curve (fig. 15) is for a heavily loaded (843.7 foot-pounds per stroke) 12-foot mill. It starts with a velocity of about 10 miles an hour, and reaches a maximum at about 23 miles, with a speed of 57 strokes per mile. At 30 miles it is making 51 strokes per mile. The maximum point of this curve is much farther to the right than that of any other curve. The number of strokes increases from 5 per minute at 12 miles to 25 per minute at 30 miles.

Mill No. 12.—This is a 14-foot Ideal, shown in Pl. VII. The tower is of steel and is 30 feet high to the axis of the wheel. The wheel has 24 curved sails, each 48¾ by 17½ by 8 inches, set at an angle of 30° with the plane of the wheel. It is back-geared, 2½ to 1. The pump is of the Frizell make (shown in fig. 16). The discharge pipe is 10 inches in diameter, the cylinder 9½ inches in diameter, the supply pipe 6 inches in diameter, terminating in a well point 10 feet long and 6 inches in diameter, the lower end of which is 32 feet below the surface of the ground. The lift, as nearly as could be estimated, was 11 feet, the discharge per stroke 11.6 quarts. The mean barometric pressure was 27.04 inches, the mean temperature 81° F. The water is pumped into a reservoir 100 feet

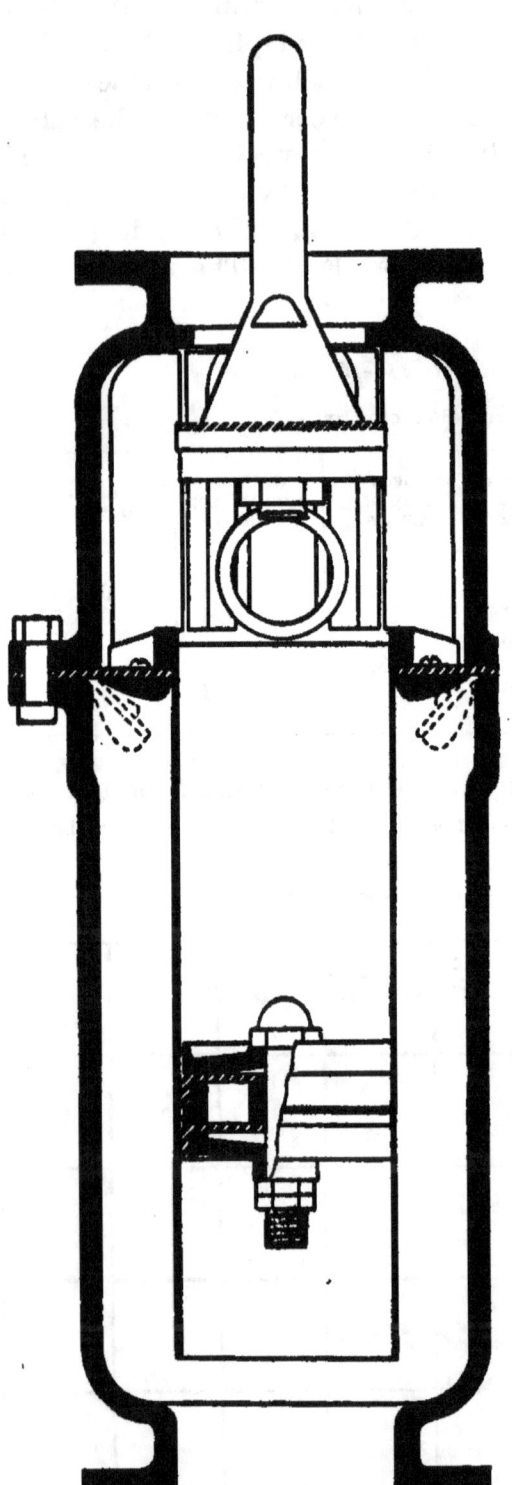

FIG. 16.—Working parts of Frizell cylinder.

by 100 feet by 3 feet deep. The pump had been in use about one year. The results of the tests are as follows:

Results of tests of mill No. 12—14-foot Ideal.

[Load per stroke, 263.5 foot-pounds.]

Wind velocity per hour.	Revolutions of wind wheel per minute.	Strokes of pump per minute.	Gallons pumped per minute.	Useful horse-power.
8 miles	7.7	3.1	8.9	0.025
12 miles	27.2	10.9	30.1	0.087
16 miles	39.3	15.7	45.1	0.125
20 miles	48.0	19.2	55.2	0.158
25 miles	53.7	21.5	61.8	0.173

This curve (fig. 17) is for a very lightly loaded (263.5 foot-pounds) 14-foot mill. This load is only 31 per cent of that of the 12-foot mill, No. 11. The curve starts in a 7-mile to an 8-mile wind, and reaches a maximum at 15 miles, with a speed of 58 strokes per mile of wind. At 30 miles the speed is 47 strokes. Although this is a very lightly

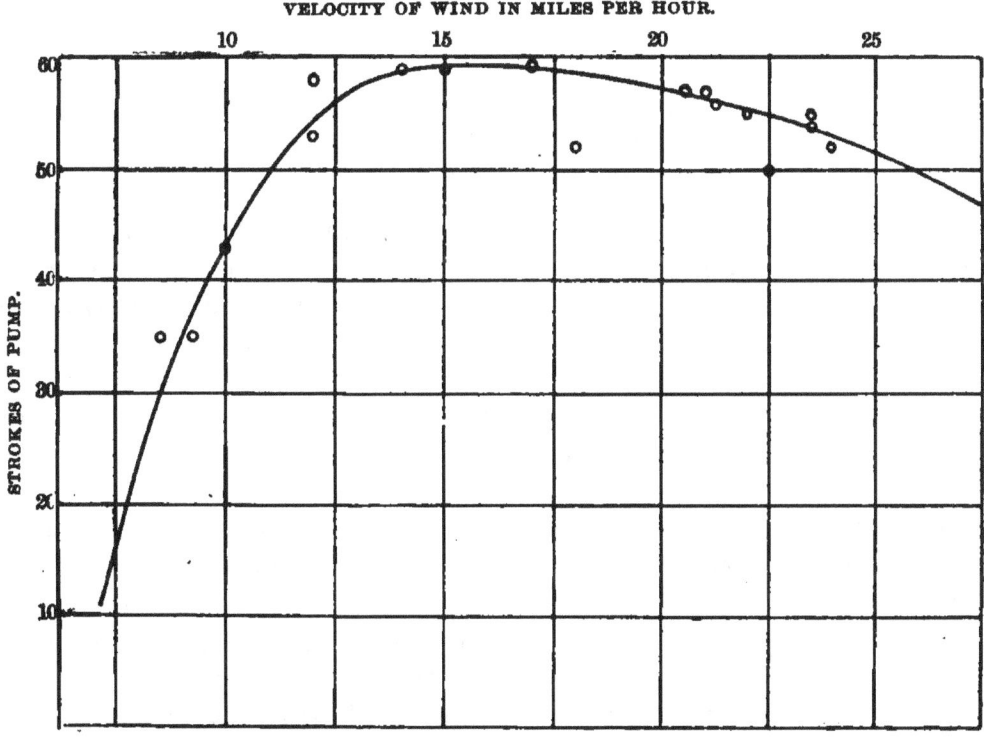

FIG. 17.—Diagram showing results with mill No. 12—14-foot Ideal.

loaded mill it does not make many strokes per minute. In fact, although it is not as heavily loaded as the 12-foot mill, No. 3, it does not make as many strokes per minute as the latter mill, and it is producing much less power than mill No. 3.

Mill No. 13.—This is a 12-foot Aermotor (shown in Pl. VIII). The tower is of wood, with the axis of the wheel 25 feet above the ground. The exposure was good and the plant in excellent condition, having been in use about one year at the time of tests. The wheel is the same as that of No. 3. The pump is of the Stone make. The discharge pipe is 10 inches in diameter, the supply pipe 5 inches in diameter, on a well point 10 feet long, the lower end of which is 17 feet below the surface of the ground. The length of stroke is 12 inches. The plunger valve is of the double-flap type, and the check valve of the single-flap type. The discharge per stroke at the time of test was 14.4 quarts and the lift about 11 feet. The mean barometric pressure was 27.09 inches, the mean temperature 91° F. The results of the tests are as follows:

Results of tests of mill No. 13—12-foot Aermotor.

[Load per stroke, 330 foot-pounds.]

Wind velocity per hour.	Revolutions of wind wheel per minute.	Strokes of pump per minute.	Gallons pumped per minute.	Useful horse-power.
12 miles	36.7	11.0	39.6	0.110
16 miles	57.0	17.1	61.6	0.171
20 miles	70.0	21.0	75.6	0.210
25 miles	82.0	24.6	88.6	0.247
30 miles	91.7	27.5	99.0	0.275

The important difference between this plant and No. 3 is that the latter has a 4-inch supply pipe and an open well, while the former has a 5-inch supply pipe on a well point. The useful load per stroke of mill No. 13 is 20 per cent less than that of mill No. 3, and the number of strokes per minute of No. 13 is slightly greater than that of No. 3. It appears that the well point offers some resistance, but how much can not be said from this data.

Mill No. 14.—This is a 12-foot Gem, like the one shown in Pl. IX, on a 60-foot steel tower. The pump is of the Gause make. The cylinder is 8 inches in diameter, the length of stroke 9 inches. The supply is from a 12-inch pipe in an open well. The discharge per stroke was 9¾ quarts and the lift 15½ feet. The wind velocity was not measured.

Mill No. 15.—This is a 10-foot Gem similar to that shown in Pl. IX. The tower is of wood, the axis of the wheel being 34 feet above the ground. The mill was in good working order, but the exposure was not good, on account of trees. The wheel has 24 sails, each 36 by 11 by 4¾ inches, set at an angle of 35° with the plane of the wheel. It is back-geared, 3 to 1. The pump is of the Stone make. The discharge pipe is 8 inches in diameter. The supply pipe is on a 3-inch well point 8 feet long, the lower end of which is 21½ feet below the surface of the ground. The plunger valve is of the single-flap form. Depth to water is 10 feet. The discharge per stroke was 7 quarts, the lift about 15 feet. The mean barometric pressure was 27.05 inches, the mean temperature 84° F.

The results of the tests are as follows:

Results of tests of mill No. 15—10-foot Gem.

[Load per stroke, 219 foot-pounds.]

Wind velocity per hour.	Revolutions of wind wheel per minute.	Strokes of pump per minute.	Gallons pumped per minute.	Useful horse-power.
12 miles	23.4	7.8	13.6	0.053
16 miles	35.7	11.9	21.0	0.082
20 miles	44.1	14.7	25.7	0.101
25 miles	43.8	14.6	25.5	0.099

This mill revolves very slowly, indicating a heavy load. Its useful horsepower, however, is little greater than that of mill No. 5.

Mill No. 16.—This is a 10-foot Halliday, pumping water into the same pond as No. 15. It is similar to the mill shown in fig. 20. The tower is of wood, the axis of the wheel being 28 feet above the ground. The wheel has 78 sails, each $36\frac{1}{4}$ by 4 by $2\frac{1}{4}$ inches, set at an angle of 35.5° to the plane of the wheel. It is not back-geared. The pump is of the Gause make, with a discharge pipe 6 inches in diameter and a supply pipe 4 inches in diameter. There is a 6-inch galvanized-iron pipe, forming an open well, extending 15 feet into the water. The depth to water was 11 feet, the lift 16 feet, and the discharge per stroke 3 quarts. The mean barometeric pressure was 27.02 inches, the mean temperature 94° F. The results of tests are as follows:

Results of tests of mill No. 16—10-foot wooden Halliday.

[Load per stroke, 100 foot-pounds.]

Wind velocity per hour.	Revolutions of wind wheel per minute.	Strokes of pump per minute.	Gallons pumped per minute.	Useful horse-power.
8 miles	4.0	4.0	3.0	0.012
12 miles	22.6	22.6	16.9	0.067
16 miles	33.9	33.9	25.4	0.103
20 miles	42.7	42.7	32.0	0.130
25 miles	52.5	52.5	39.3	0.159

This 10-foot wooden mill is seen to be doing more useful work than the 10-foot steel Gem, No. 15. Usually, however, the direct-stroke wooden mills do less work than the back-geared steel mills of the same size.

Mill No. 17.—This is a 12-foot improved Gem on a 30-foot steel tower. The wheel has 32 curved sails, each 42 by $11\frac{1}{2}$ by $4\frac{3}{4}$ inches, set at an angle of 37° with the plane of the wheel. It is back-geared, 2 to 1. The pump is of the Gause type, with an 8-inch discharge pipe, a 4-inch supply pipe, 12 inches stroke, and an open well formed of a 12-inch wooden casing. The depth to water was $17\frac{1}{4}$ feet, and the

discharge per stroke 8¾ quarts. The mean barometric pressure was
27.05 inches, the mean temperature 93° F. The results of the tests
are as follows:

Results of tests of mill No. 17—12-foot improved Gem.

[Load per stroke, 385 foot-pounds.]

Wind velocity per hour.	Revolutions of wind wheel per minute.	Strokes of pump per minute.	Gallons pumped per minute.	Useful horse-power.
12 miles _____	12.0	6.0	12.7	0.070
16 miles_____	25.6	12.8	27.2	0.149
20 miles_____	34.6	17.3	36.8	0.202

This mill, although nearly new, does not work well. It is out of
plumb. Only a few measurements of the number of strokes per mile
of wind were made.

Mill No. 18.—This is an 8-foot Ideal on a 36-foot wooden tower.
The exposure was good and the parts in good working order. The
wheel is like that of mill No. 4. The pump is of the Stone make, with
a 6-inch discharge pipe. There is no supply pipe, the cylinder being
under water, with 3 inches opening to it from below. The check valve
is of the lift variety, the plunger valve of the single-flap variety. The
well is dug to a depth of 8 feet. It is 4½ feet in diameter. In the
bottom a 10-inch galvanized-iron pipe extends down several feet. It
was 11 feet to water. The lift was 14½ feet and the discharge per
stroke 2.92 quarts. The mean barometric pressure was 27.01 inches,
the mean temperature 83° F. The cost of the plant, including pond,
was $125. The results of the tests are as follows:

Results of tests of mill No. 18—8-foot Ideal.

[Load per stroke, 89.2 foot-pounds.]

Wind velocity per hour.	Revolutions of wind wheel per minute.	Strokes of pump per minute.	Gallons pumped per minute.	Useful horse-power.
8 miles_____	20.0	8.0	5.8	0.022
12 miles_____	50.5	20.2	14.6	0.054
16 miles_____	65.2	26.1	18.9	0.070
20 miles_____	70.0	28.0	20.3	0.076
25 miles_____	68.7	27.5	19.9	0.074

After the spring which holds the wind wheel of this mill in the wind
was tightened, the number of strokes per mile of wind was increased
from 98 to 114 in a 16-mile wind, from 84 to 102 in a 20-mile wind, and
from 66 to 79 in a 25-mile wind.

Mill No. 19.—This is a 12-foot Gem (shown in Pl. IX) on a 30-foot
wooden tower. The exposure was good and the mill in good working order. The wheel is like that of mill No. 17. The pump is of
the Stone make, with a 10-inch discharge pipe and a 4-inch supply

pipe. The length of the stroke is 10 inches. The supply pipe is on a 4-inch well point 9 feet long, the end of which is 23 feet below the surface of the ground. The check valve is of the lift type and the plunger valve of the single-flap type. The lift was about 18 feet and the discharge per stroke 12 quarts. The mean barometric pressure was 27.13 inches, the mean temperature 70° F. The water is pumped into a reservoir 120 feet by 60 feet. The results of the tests are as follows:

Results of tests of mill No. 19—12-foot Gem.

[Load per stroke, 450 foot-pounds.]

Wind velocity per hour.	Revolutions of wind wheel per minute.	Stroke of pump per minute.	Gallons pumped per minute.	Useful horse-power.
12 miles	12.4	6.2	18.6	0.085
16 miles	23.8	11.9	35.7	0.162
20 miles	29.4	14.7	44.1	0.201
25 miles	32.0	16.0	48.0	0.219

This curve (fig. 18) shows that a 9-mile wind is necessary to start this mill, and that the greatest number of strokes per mile of wind

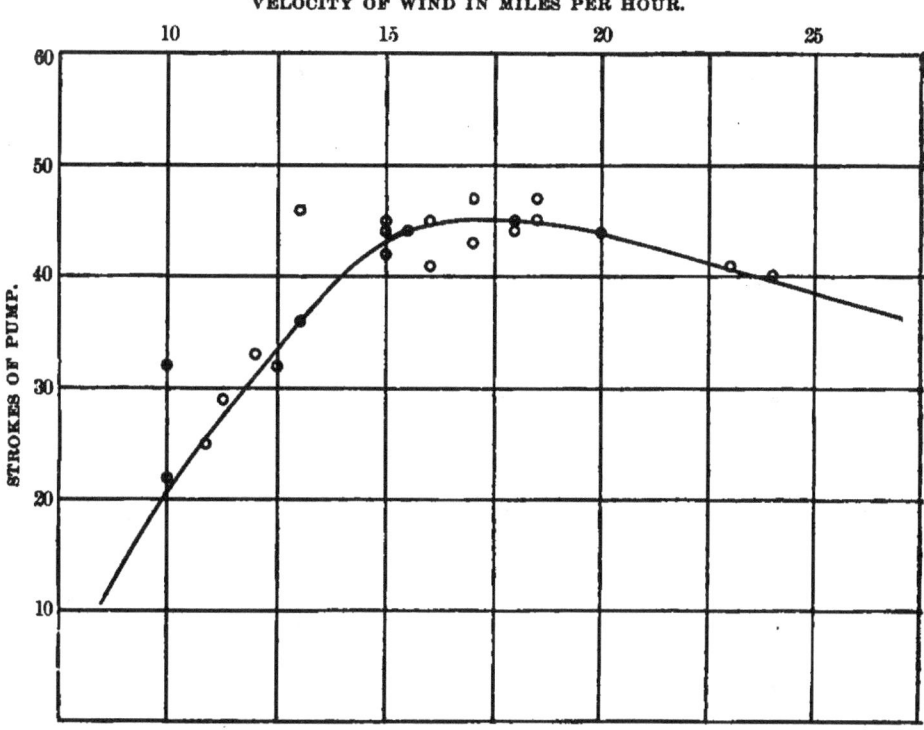

FIG. 18.—Diagram showing results with mill No. 19—12-foot Gem.

is 45—about 25 per cent less than most 12-foot steel back-geared mills, being at the rate of 6 strokes at 12 miles and 16 strokes at 25 miles an hour. The load of this mill is somewhat greater than that of No. 3, and its power should be equal or greater, but it is seen to be much less.

Mill No. 20.—This is a 15½-foot Jumbo (shown in Pl. X, *A*). Its axis is a steel shaft 8 feet above the ground. It has 6 sails, each 9½ by 3¼ feet, with an outer radius of 7¾ feet and an inner radius of 4½ feet. This mill operates two pumps, one at each end of the axis, each pump having a 6-inch cylinder, a 3-inch discharge pipe, and a 3-inch supply pipe on a well point 5 feet long. The discharge per stroke of the two pumps was 10 quarts, the lift about 14 feet. The mean barometric pressure was 27.09 inches, the mean temperature 85° F. The anemometer was held 14 feet above the surface of the ground, or at the elevation of the center of a sail when in its highest position. The wheel is set in a large box the top of which is on a level with the axis of the wheel, to prevent the wind from striking the part of the wheel below its axis. The results of the tests are as follows:

Results of tests of mill No. 20—15½-foot Jumbo.

[Load per stroke, 291.2 foot-pounds.]

Wind velocity per hour.	Revolutions of wind wheel per minute.	Strokes of pump per minute.	Gallons pumped per minute.	Useful horse-power.
16 miles	5.3	5.3	13.2	0.047
20 miles	10.7	10.7	26.7	0.095

The curve shown in fig. 19 is seen to be different from the others in that it apparently has no maximum. It starts at a wind velocity of 13

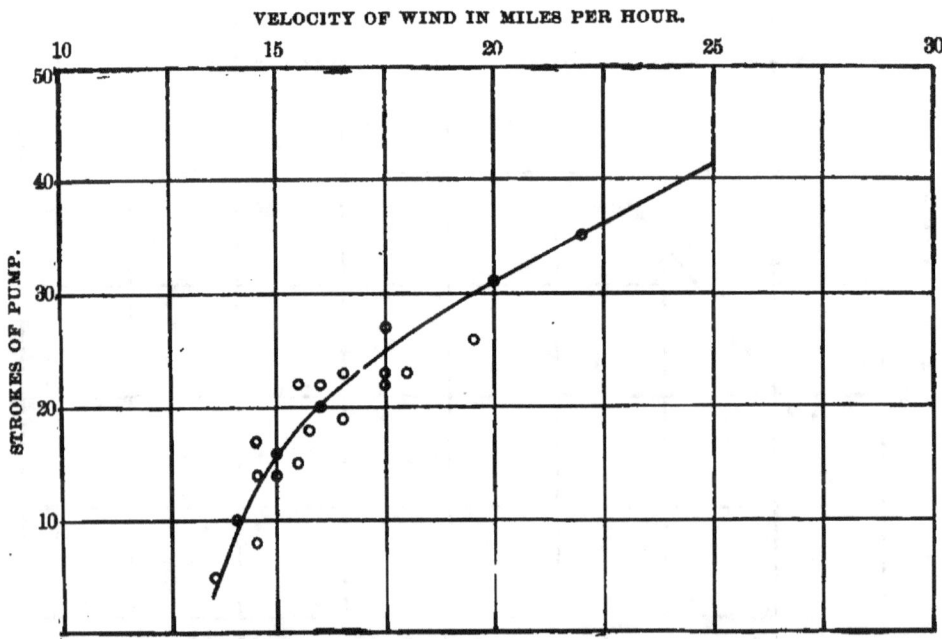

FIG. 19.—Diagram showing results with mill No. 20—15½-foot Jumbo.

to 14 miles, and beyond 20 miles it appears to be a nearly straight line. Comparing these results with those of No. 5, it will be seen that this mill is inferior in power to a good 8-foot steel mill.

Mill No. 21.—This is a 12-foot Halliday (shown in Pl. XI) on a 31-foot wooden tower.   It was made by the United States Wind Engine and Pump Company, of Batavia, Illinois.   The working parts are

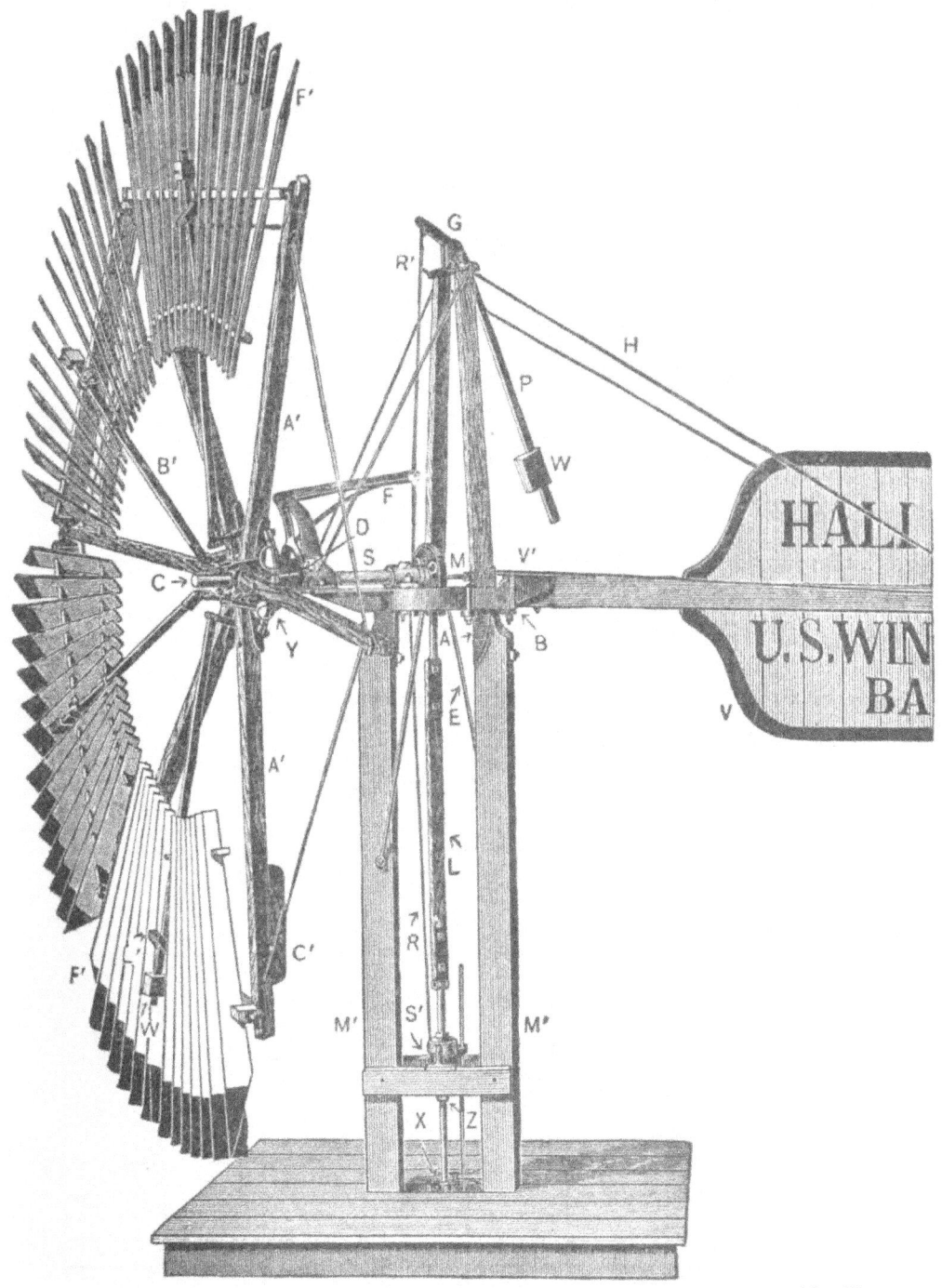

FIG. 20.—Working parts of Halliday mill: *A*, bed plate; *A'*, fan arm; *B*, turntable; *B'*, regulating rod; *C*, front plate; *C'*, assisting weight; *D*, sliding head; *E*, tie rod; *F*, forked lever; *F'*, fans or sails; *G*, truss frame; *H*, truss rod; *L*, pitman; *L'*, fan lever; *M*, crank plate; *M' M'*, masts; *P*, weight lever; *R*, shut-off rod; *R'*, shut-off rod lever; *S*, main shaft; *S'*, sleeve; *V*, vane; *V'*, vane arm; *W*, weight; *W'*, counterweight; *X*, swivel box; *Y*, spider; *Z*, sliding box.

shown in fig. 20.   The wheel has 64 sails, each 42½ by 5 by 2⅝ inches, set at an angle of 35° with the plane of the wheel.   It is not back-geared, and regulates itself on the centrifugal principle—the sails taking the direction of the wind. The pump is of the Stone make, with 7½-inch discharge pipe, 4-inch supply pipe, and 7 inches stroke. The check valve is of the lift variety, the plunger valve of the double-flap variety. The well, which is open, is formed by a wooden curb, 12 inches in diameter, sunk in the bottom of a dug well 9 feet deep. The depth to water was 11½ feet and the lift 11 feet. The discharge per stroke was 4½ quarts when pumping quite rapidly (30 strokes per minute). The valves were not in very good repair and the pump lost its priming after a time. The results of the tests are as follows:

Results of tests of mill No. 21—12-foot wooden Halliday.

[Load per stroke, 141.5 foot-pounds.]

Wind velocity per hour.	Revolutions of wind wheel per minute.	Strokes of pump per minute.	Gallons pumped per minute.	Useful horse-power.
12 miles	14.0	14.0	15.9	0.060
16 miles	28.5	28.5	32.1	0.121
20 miles	37.3	37.3	42.0	0.159
25 miles	44.6	44.6	50.2	0.184

This curve (shown in fig. 21), which is for a lightly loaded (141.5 foot-pounds) direct-stroke 12-foot mill, will be seen to start at a wind

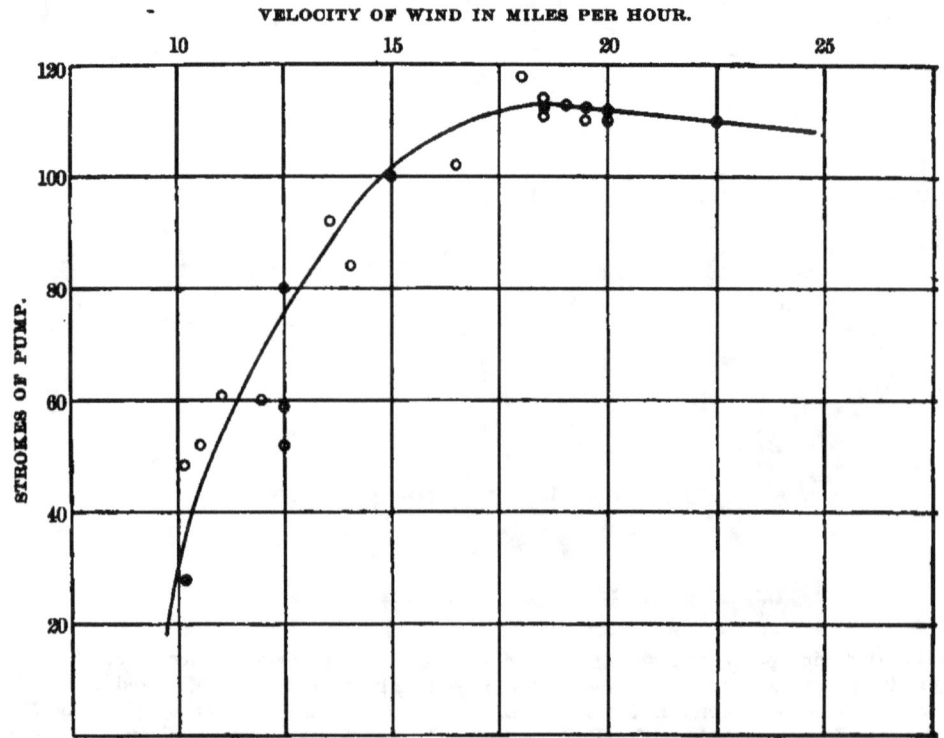

FIG. 21.—Diagram showing results with mill No. 21—12-foot Halliday.

velocity of 9 to 10 miles and to reach a maximum at 19 miles, with a speed of 112 strokes per mile. At 30 miles the number of strokes is about 98 per mile. The number of strokes per minute varies from 14 at 12 miles to 45 at 25 miles. The power of this mill is only about half that of the 12-foot Aermotor, No. 3.

Mill No. 25.—This is an 8-foot steel mill on a 32-foot steel tower, made by Fairbanks, Morse & Company. The wheel has 18 curved sails, each 29 by 11⅝ by 5¼ inches, set at an angle of 29° with the plane of the wheel. It is back-geared, 2½ to 1. The pump is of the common hand variety, with a 2½-inch cylinder, 1½-inch supply and discharge pipes, and 4 inches stroke. The well is open, 6¼ feet to water. The discharge per stroke was 0.31 quart and the lift 8½ feet. The water raised is used for watering stock. The results of the tests are as follows:

Results of tests of mill No. 25—8-foot Fairbanks-Morse steel mill.

[Load per stroke, 5.5 foot-pounds.]

Wind velocity per hour.	Revolutions of wind wheel per minute.	Strokes of pump per minute.	Gallons pumped per minute.	Useful horse-power.
8 miles	30.7	12.3	1.0	0.002
12 miles	73.0	29.2	2.3	0.005
16 miles	93.2	37.3	2.9	0.007
20 miles	95.0	38.0	3.0	0.008
25 miles	83.2	33.3	2.6	0.005

This is a very lightly loaded mill—only 5.5 foot-pounds per stroke of pump. The number of strokes per minute in light winds is large; the number of strokes per minute in a 25-mile wind is only 33.3, compared with 38 in a 20-mile wind. A comparison of this mill with No. 5 will show the difference between a small pumping outfit for stock purposes and one for irrigation.

Mill No. 30.—This is a 16-foot Irrigator manufactured by M. Schow, of Kinsley, Kansas. It is a power mill, but not geared forward, and works a pump called a "water elevator." The tower is of wood, 22 feet to axis of wheel. The wind wheel has 10 plane wooden sails, each 70½ by 16 by 13½ inches, set at an angle of 39° to the direction of the wind.

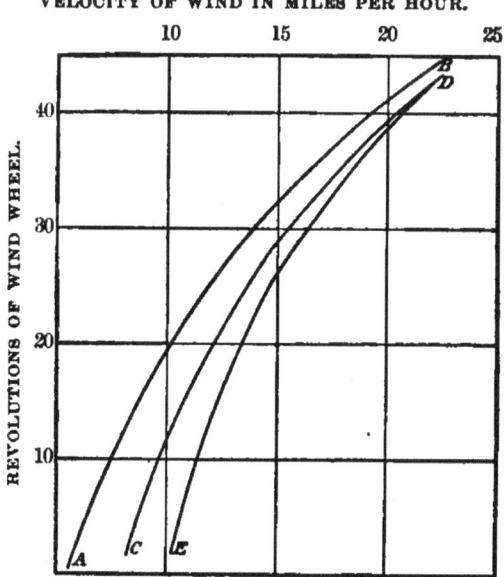

Fig. 22.—Diagram showing revolutions of wind wheel of mill No. 30—16-foot Irrigator. *AB* is for a load of 32 foot-pounds per revolution of wind wheel; *CD* is for a load of 251 foot-pounds; *ED* is for a useful pump load of 337 foot-pounds.

The vertical shafting is geared back 30 to 13, and the horizontal shaft is geared forward 13 to 30. The water is lifted from a well 8 feet

by 10 feet (7 feet depth to water), cased with wood.   The buckets
of the elevator are made of galvanized iron and are 14 by 7½ by 5½
inches, set 18 inches center to center, and hold 3 gallons each.   Each
bucket has a valve in its bottom.   The lift is about 11 feet, and there
are 20 buckets on the elevator chains.   There is a box in the bottom
of the well into which the buckets dip to get their supply of water.
This box has a screen in its bottom to keep out the sand.   The mill
does not govern well, on account of side draft.   The mean barometric
pressure was 27.7 inches, the mean temperature 72° F.   The results
of the tests are as follows:

Results of tests of mill No. 30—16-foot Irrigator.

Load on brake.	Load per revolution of wind wheel.	Number of revolutions of wind wheel per minute at given wind velocities (per hour).					Useful horsepower at given wind velocities (per hour).				
		8 miles.	12 miles.	15 miles.	20 miles.	25 miles.	8 miles.	12 miles.	15 miles.	20 miles.	25 miles.
Pounds.	*Ft.-lbs.*										
2	32	12	26	33	41	44	0.012	0.025	0.032	0.040	0.043
16	251	....	19	29	39	42	....	0.140	0.220	0.300	0.320
Pump.	337	....	14	26	38	42	....	0.140	0.250	0.400	0.440

Fig. 22 shows the number of revolutions per minute of the wind
wheel for three loads.   Fig. 23 shows the horsepower for these loads.

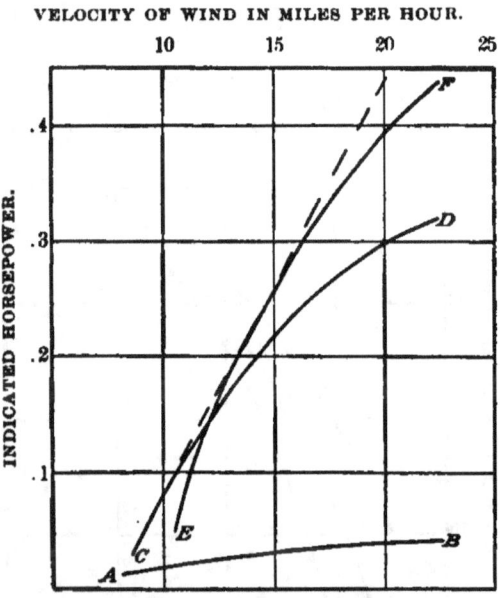

VELOCITY OF WIND IN MILES PER HOUR.

FIG. 23.—Diagram showing horsepower of mill
No. 30. Curve *AB* shows the horsepower for a
32 foot-pound load; *CD*, the power for a 251 foot-
pound load; *EF*, the power for a 337 foot-pound
load; dotted curve shows maximum horsepower
for best load.

Comparing the useful pump
horsepower of this mill with
that of the 16-foot mill No. 9,
it will be seen that this mill is
not so powerful as No. 9.   The
power of the Aermotor for low
velocities is much greater than
that of the Irrigator.

Mill No. 31.—This is a 14-foot
Elgin wooden power mill, manufactured at Elgin, Illinois.   It
works a rotary (Wonder) pump.
The wind wheel is on a 48-foot
wooden tower.   It has 88 plane
wooden sails, each 52 by 6 by
3⅛ inches, set at an angle of 37°
to the plane of the wheel.   It
is a sectional vaneless wheel;
in place of the vane there is a
heavy counterpoise.   The wheel
is geared forward about 7.19 to

1.   The rotary pump is a 3 inch.   It is on a well point 6 inches in
diameter and 8 feet long; the point penetrates the water to a depth of
8 feet.   The lift was about 18 feet.   The suction and discharge pipes

are each 3 inches in diameter. The discharge was 2 quarts per revolution of pump. The pump is manufactured by the National Pump Company, of Kansas City, Missouri. In a 15-mile to a 20-mile wind the windmill worked the pump before priming, but would not start it in that wind after priming. After several attempts of the mill to start the pump the pulley turned on the shaft so that it could not be used. It was very evident that the pump was too great a load for the mill. The owner stated that the mill would only run during a strong wind.

Mill No. 32.—This mill is like the 12-foot Aermotor, the working parts of which are shown in fig. 7. It is on a 40-foot steel tower. The wheel has 18 curved sails, each 44 by 18¾ by 7¾ inches, set at an angle of 31° to the plane of the wheel. The pump is of the Woodmanse type, a sectional view of which is shown in fig. 5, and has an 8-inch cylinder and 12 inches stroke. It is in an open well. The depth to water was 14 feet, the lift 20 feet, and the discharge per stroke 7 quarts. It is back-geared, 3⅓ to 1. The water is pumped into a pond and used for irrigation. The mean barometric pressure was 27.9 inches, the mean temperature 82° F. The results of the tests are as follows:

Results of tests of mill No. 32—12-foot Aermotor.

[Load per stroke, 313 foot-pounds.]

Wind velocity per hour.	Revolutions of wind wheel per minute.	Strokes of pump per minute.	Gallons pumped per minute.	Useful horse-power.
8 miles................	17.7	5.3	9.3	0.047
12 miles................	49.3	14.8	25.9	0.131
15 miles................	60.0	18.5	32.4	0.164
20 miles................	73.0	22.0	38.5	0.195

Mill No. 33—This is a 10-foot Woodmanse pumping mill on a 40-foot steel tower. The working parts are like those shown in fig. 4. The wheel has 24 curved sails, each 30½ by 12½ by 5½ inches, set at an angle of 29° to the plane of the wheel. It is back-geared, 2½ to 1. The pump is of the Woodmanse make (like that shown in fig. 5), with a 6-inch cylinder and 10 inches stroke. The depth to water was 14 feet. The pump is on a well point 3¼ inches in diameter, 4 feet long, and 67 feet below the surface of the ground. The suction pipe is 3¼ inches in diameter, the discharge pipe 6 inches in diameter. The lift was about 22 feet, the discharge per stroke 3¾ quarts. The water is pumped into a pond and used for irrigation. The mean barometric pressure was 27.9 inches, the mean temperature 88° F.

The results of the tests are as follows:

Results of tests of mill No. 33—10-foot Woodmanse.

[Load per stroke, 173 foot-pounds.]

Wind velocity per hour.	Revolutions of wind wheel per minute.	Strokes of pump per minute.	Gallons pumped per minute.	Useful horse-power.
8 miles...............	8.7	3.5	3.3	0.020
12 miles...............	36.0	14.4	13.5	0.075
15 miles...............	51.2	20.5	19.2	0.110

Mill No. 35.—This is an 8-foot steel Dempster pumping mill, manufactured by the Dempster Manufacturing Company, of Beatrice, Nebraska, and is shown in Pl. XII. The tower is of steel, 30 feet to axis of wheel. The wind wheel has 18 curved sails, each 30½ by 14½ by 7 inches, set at an angle of 27° to the plane of the wheel. It is back-geared, 3 to 1. The well is an open dug well. The distance from the surface of the ground to the water was 39 feet. The water is pumped into a tank 22 feet above ground, and is used for irrigation. The pump has an 8-inch stroke, a 3¼-inch cylinder, and 2-inch suction and discharge pipes. The lift was 58 feet, the discharge per stroke 1.1 quarts. The mean temperature was 51° F., the mean barometric pressure 28.9 inches. The results of the tests are as follows:

Results of tests of mill No. 35—8-foot steel Dempster.

[Load per stroke, 133.5 foot-pounds.]

Wind velocity per hour.	Revolutions of wind wheel per minute.	Strokes of pump per minute.	Gallons pumped per minute.	Useful horse-power.
12 miles...............	57.6	19.2	5.3	0.077
16 miles...............	86.4	28.8	7.9	0.116
20 miles...............	99.9	33.3	9.2	0.134

In a 15-mile wind the pump made 26 to 27 strokes per minute, and when the pump rod was uncoupled from the pump it made 34 strokes per minute in the same wind. Fig. 24 shows the number of strokes per minute for different wind velocities. This mill is heavily loaded—133 foot-pounds per stroke of pump. It starts at a wind velocity of about 9 miles an hour, and as the wind increases the number of strokes increases rapidly at first, then more slowly. At 25 miles an hour the number of strokes is 37 per minute. This mill is back-geared, 3 to 1, so that in a 25-mile wind the wind wheel makes 111 revolutions per minute and has a circumference velocity of 46.5 feet per second.

Mill No. 36.—This is a 22½-foot Eclipse wooden pumping mill, lifting water into a tank for railroad purposes. The tower is of wood, 52 feet to the axis of the wheel. The wind wheel has 136 plane sails, each 105 by 6 by 2 inches, set at an angle of 39° to the plane of the wheel. It works by direct stroke. The well, which is open, is 20 feet in diameter; the depth to water was 19 feet from the surface of the ground. The pump is double-acting, and has a 4½-inch cylinder, 7 inches stroke, and 2-inch suction and discharge pipes. The lift was 39 feet. The tank is 90 feet from the well. The discharge could not be measured, but as the packing of the pump was new it was approximately equal to twice the volume of the

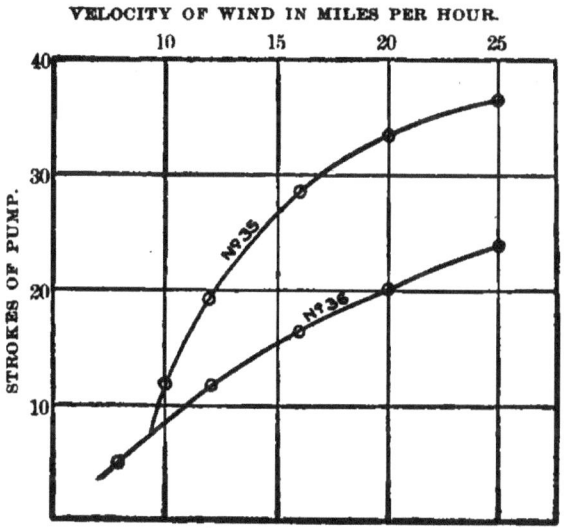

FIG. 24.—Comparative diagram showing results with mills No. 35 (8-foot Dempster) and No. 36 (22½-foot Eclipse).

cylinder, or 0.76 gallon per double stroke. The mean temperature was 58° F., the mean barometric pressure 29.43 inches. The results of the tests are as follows:

Results of tests of mill No. 36—22½-foot wooden Eclipse.

[Load per stroke, 248 foot-pounds.]

Wind velocity per hour.	Revolutions of wind wheel per minute.	Strokes of pump per minute.	Gallons pumped per minute.	Useful horse-power.
8 miles.............	4.8	4.8	3.8	0.036
12 miles.............	12.0	12.0	9.4	0.090
16 miles.............	16.5	16.5	12.8	0.124
20 miles.............	20.0	20.0	15.6	0.150
25 miles.............	24.2	24.2	18.8	0.182

Fig. 24 shows the strokes per minute at different wind velocities. The diagram is seen to be quite different from that of mill No. 35. The mill starts at a wind velocity of 6 or 7 miles an hour, and increases gradually to 24 strokes in a 25-mile wind. The circumference velocity of the latter is 46.5 feet per second, that of the former 29 feet. The wind wheel of this mill is making 24 revolutions per minute; that of No. 35 is making 111 revolutions per minute in a 25-mile wind.

Mill No. 37.—This is a 12-foot steel Woodmanse Mogul like that shown in fig. 4. It pumps water into a pressure tank 9½ feet by 2½ feet in a cellar and about 170 feet from the well. The pump and pressure tank are shown in fig. 25.   The tower is of steel, and is 40 feet to the axis of the wheel.   The wind wheel has 30 curved sails, each 37 by 13 by 5.5 inches, set at an angle of 29° to the plane of the wheel.   The well is 50 feet deep and 8 inches in diameter; the depth to water was 42 feet.   The pump has a 3-inch cylinder and 9 inches stroke.   The discharge pipe is 1 inch in diameter, the suction pipe 1¼ inches in

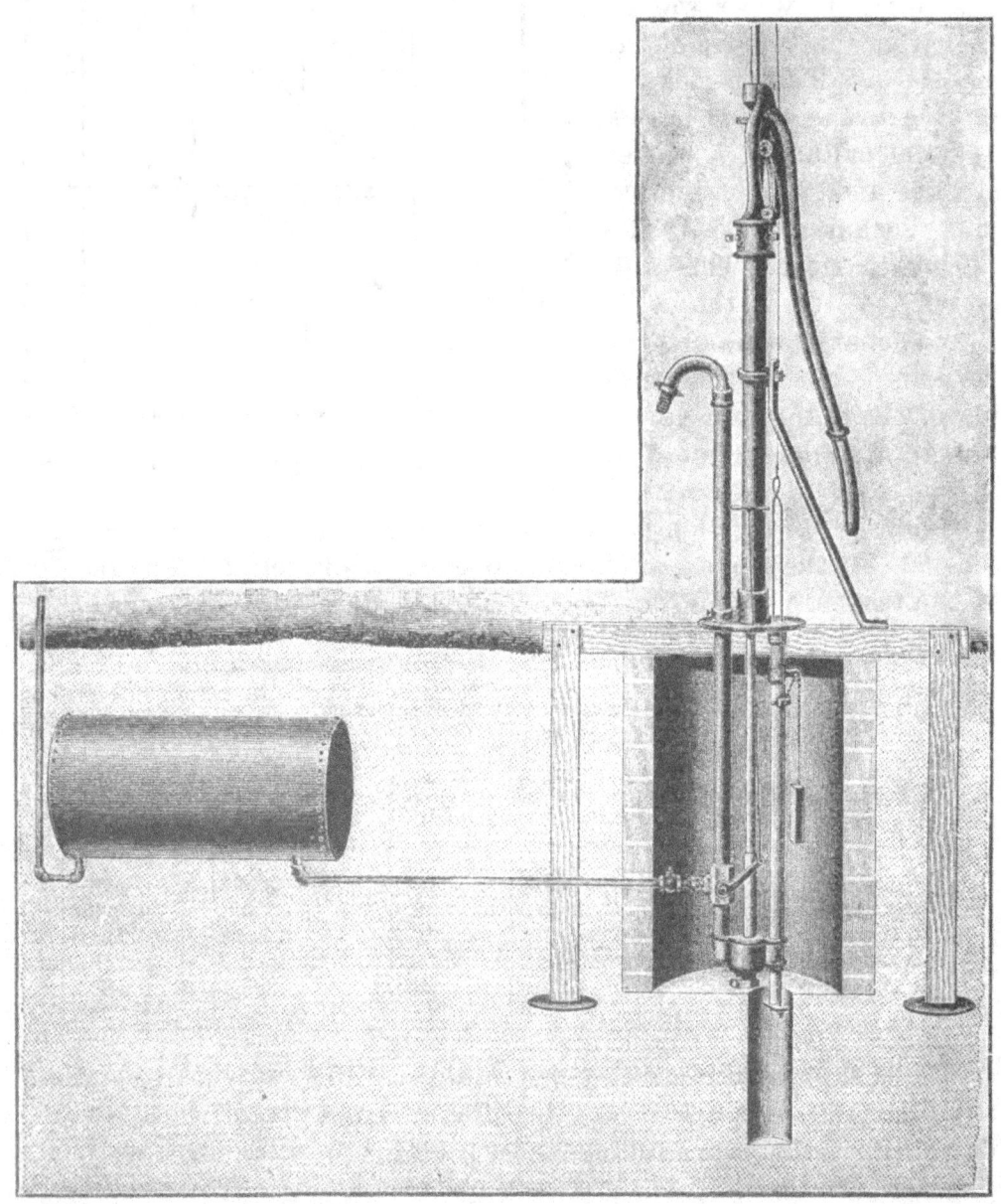

FIG. 25.—Pump, pressure tank, and hydraulic regulator of mill No. 37—12-foot Woodmanse Mogul.

diameter.   The discharge per stroke was 1.05 quarts.   The load on the pump was equal to 43 feet head of water (the friction in about 200 feet of 1-inch pipe) and a compressed-air pressure the amount of which was recorded on a gage.   The mean temperature was 66° F., the mean barometric pressure 29.1 inches.   The results of the tests are as follows:

Results of tests of mill No. 37—12-foot steel Woodmanse Mogul.

Load per stroke.	Number of strokes of pump per minute at given wind velocities (per hour).						Useful horsepower at given wind velocities (per hour).					
	8 miles.	12 miles.	16 miles.	20 miles.	25 miles.	30 miles.	8 miles.	12 miles.	16 miles.	20 miles.	25 miles.	30 miles.
Ft.-lbs.												
94	7.7	18.4	24.5	29.3	34.2	36.0	0.022	0.053	0.071	0.083	0.097	0.108
254	......	12.2	19.0	23.3	26.7	29.0	......	0.094	0.146	0.180	0.205	0.223
350	......	......	......	18.0	21.0	21.5	......	......	......	0.191	0.223	0.228
473	......	......	......	10.7	......	......	......	......	......	0.153	......	......

Fig. 26 shows the number of strokes per minute for different wind velocities for four useful pump loads, viz, 43 feet, 43 feet plus 32 pounds, 43 feet plus 50 pounds, and 43 feet plus 75 pounds, or, reducing the pounds pressure to head in feet, the four useful loads are: 43 feet for the curve aa', 116 feet for the curve bb', 160 feet for the curve cc', and 216 feet for the curve dd'. The effect of increased load on the number of strokes is well shown here. The effect of the hydraulic regulator may be seen in the curve cc', but it is shown to a greater extent in the curve dd'.

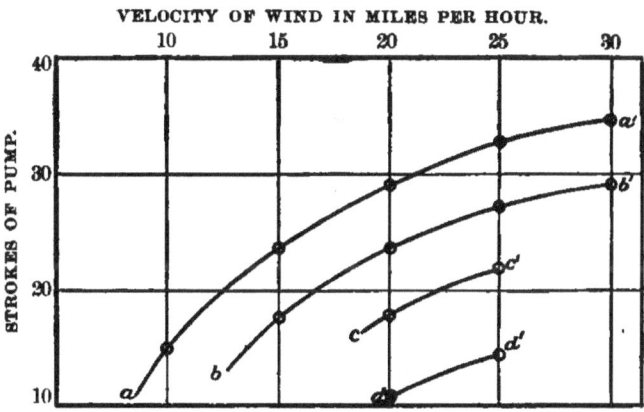

VELOCITY OF WIND IN MILES PER HOUR.

FIG. 26.—Diagram showing results with mill No. 37—12-foot Woodmanse Mogul. The curve aa' is for a useful pump load of 43 feet head; bb' is for a useful pump load of 116 feet head; cc' is for a useful pump load of 160 feet head; dd' is for a useful pump load of 216 feet head. The effect of the hydraulic regulator is shown in the curves cc' and dd'.

Mill No. 38.—This is a 10-foot Woodmanse wooden mill on a 30-foot wooden tower, and is used to pump water for stock. The wind wheel has 96 plane sails, each 34 by 3½ by 1¼ inches, set at an angle of 38° to the plane of the wheel. The well is driven and is 107 feet deep; the depth to water was about 44 feet. The pump works direct and has a 2½-inch cylinder and 4 inches stroke. The lift was about 50 feet, and the discharge per stroke of pump $\frac{1}{20}$ gallon.

The results of the tests are as follows:

Results of tests of mill No. 38—10-foot wooden Woodmanse.

[Load per stroke, 21 foot-pounds.]

Wind velocity per hour.	Revolutions of wind wheel per minute.	Strokes of pump per minute.	Gallons pumped per minute.	Useful horse-power.
8 miles	18	18	0.9	0.012
12 miles...............	29	29	1.5	0.019
16 miles..............	36	36	1.8	0.023
20 miles..............	41	41	2.1	0.026
25 miles..............	46	46	2.3	0.029

Fig. 27 shows the number of strokes per minute for different wind velocities. This is a lightly loaded mill, working direct stroke. It

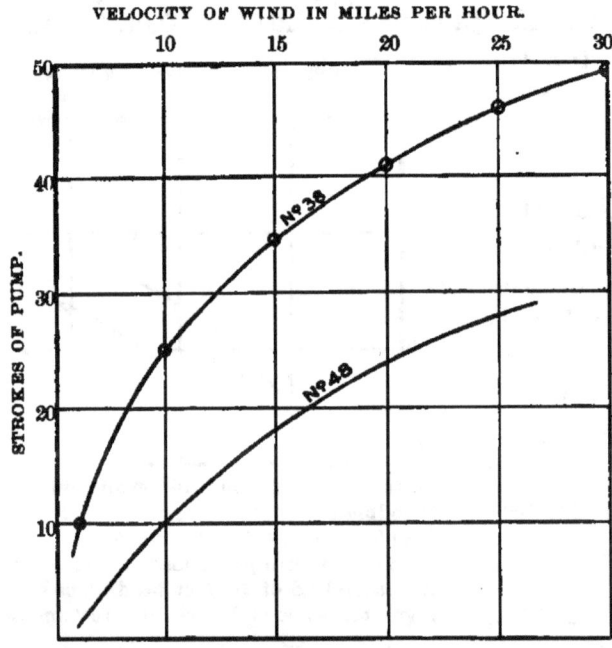

FIG. 27.—Comparative diagram showing results with mills No. 38 (10-foot wooden Woodmanse) and No. 48 (30-foot wooden Halliday).

starts in a 5-mile wind; the number of strokes increases very rapidly at first, and more slowly for high velocities. In a 30-mile wind the number of strokes per minute is 50. The circumference velocity of this wheel in a 25-mile wind is 24 feet per second.

Mill No. 39.—This is a 10-foot Woodmanse direct-stroke iron pumping mill, used to pump water for stock. The tower is of iron, 35 feet to the axis of the wheel. The wind wheel has 18 curved sails, each 30 by 12½ by 7¾ inches, set at an angle of 40° to the plane of the wheel. The pump is on a well point 24 feet below the surface of the ground. It has a stroke of 6 inches, a 3½-inch cylinder, and 1½-inch suction and discharge pipes. The lift was about 27 feet, the discharge per stroke 1 quart. The results of the tests are as follows:

Results of tests of mill No. 39—10-foot iron Woodmanse.

[Load per stroke, 36 foot-pounds.]

Wind velocity per hour.	Revolutions of wind wheel per minute.	Strokes of pump per minute.	Gallons pumped per minute.	Useful horse-power.
8 miles	15	15	3.7	0.017
12 miles	30	30	7.5	0.034
16 miles	40	40	10.0	0.044
20 miles	47	47	11.7	0.052

Mill No. 40.—This is an 8-foot steel pumping mill manufactured by Fairbanks, Morse & Company. It is used to pump water for stock. It is on a 30-foot steel tower. The wind wheel has 18 curved sails, each 29 by 10½ by 5 inches, set at an angle of 29° to the plane of the wheel. It is back-geared, 2½ to 1. The pump is on a well point 20 feet below the surface of the ground. It has a stroke of 6 inches, a 3½-inch cylinder, and 1½-inch suction and discharge pipes. The lift was about 22 feet, and the discharge per stroke $\frac{1}{15}$ gallon. The results of the tests are as follows:

Results of tests of mill No. 40—8-foot Fairbanks-Morse steel mill.

[Load per stroke, 13 foot-pounds.]

Wind velocity per hour.	Revolutions of wind wheel per minute.	Strokes of pump per minute.	Gallons pumped per minute.	Useful horse-power.
8 miles	32.5	13.0	0.8	0.005
12 miles	55.0	22.0	1.5	0.009
16 miles	72.0	29.0	1.9	0.012
20 miles	85.0	34.0	2.3	0.014

Mill No. 41.—This is a 12-foot Woodmanse direct-stroke iron pumping mill, used for pumping water for stock. The tower is of iron, 30 feet to the axis of the wheel. The pump has a 3-inch cylinder and 6 inches stroke. The wind wheel has 24 curved sails, each 36 by 14½ by 7¾ inches, set at an angle of 39° to the plane of the wheel. The lift was 28 feet, and the discharge per stroke ⅛ gallon. The results of the tests are as follows:

Results of tests of mill No. 41—12-foot iron Woodmanse.

[Load per stroke, 29 foot-pounds.]

Wind velocity per hour.	Revolutions of wind wheel per minute.	Strokes of pump per minute.	Gallons pumped per minute.	Useful horse-power.
8 miles	14	14	1.7	0.012
12 miles	25	25	3.1	0.022
16 miles	33	33	4.1	0.029
20 miles	40	40	5.0	0.036

Mill No. 42.—This is a 6-foot Ideal pumping mill on a 22-foot steel tower, and is used for pumping water for stock. It is shown in Pl. XIII. The wheel has 12 curved sails, each 26 by 14½ by 5 inches, set at an angle of 39° to the plane of the wheel. It is back-geared, 4 to 1. The well is a dug well; depth to water, 15 feet. The pump has a 3-inch cylinder and 6 inches stroke. The lift was 16 feet, the discharge per stroke 1 quart. The results of the tests are as follows:

Results of tests of mill No. 42—6-foot Ideal.

[Load per stroke, 35 foot-pounds.]

Wind velocity per hour.	Revolutions of wind wheel per minute.	Strokes of pump per minute.	Gallons pumped per minute.	Useful horse-power.
8 miles.............	16	4	1.0	0.004
12 miles.............	52	13	3.2	0.014
16 miles.............	76	19	4.7	0.020
20 miles	92	23	5.7	0.024

FIG. 28.—View of mill No. 43—10-foot Perkins.

This is the smallest mill yet tested. It is rather heavily loaded for its size; in fact, it is more heavily loaded than the 12-foot mill No. 41. It is doing very good work for a mill of its size.

Mill No. 43.—This is a 10-foot Perkins direct-stroke pumping mill on a 32-foot steel tower, and is used to pump water for stock. The working parts are shown in fig. 28. The wind wheel has 30 curved sails, each 30 by 10½ by 5 inches, set at an angle of 29° to the plane of the wheel. The pump is on a well point 34 feet below the surface of the ground. It has a 3-inch cylinder and 4 inches stroke. The lift was about 36 feet, the discharge per stroke ½ quart. The exposure was not good. The results of the tests are as follows:

Results of tests of mill No. 43—10-foot Perkins.

[Load per stroke, 37 foot-pounds.]

Wind velocity per hour.	Revolutions of wind wheel per minute.	Strokes of pump per minute.	Gallons pumped per minute.	Useful horse-power.
8 miles	20	20	2.5	0.022
12 miles	32	32	4.0	0.035
16 miles	41	41	5.1	0.045
20 miles	45	45	5.6	0.050

Mill No. 45.—This is a 10-foot Eclipse direct-stroke pumping mill used to pump water for stock. The tower is of wood, 40 feet to the axis of the wheel. The wind wheel has 84 plane sails, each 36½ by 4 by 1½ inches, set at an angle of 35° to the plane of the wheel. The well is a dug well; depth to water, 12 feet. The pump has a stroke of 6 inches, a 3-inch cylinder, and 1½-inch suction and discharge pipes. The lift was 14 feet, and the discharge per stroke 0.62 quart. The results of the tests are as follows:

Results of tests of mill No. 45—10-foot wooden Eclipse.

[Load per stroke, 18 foot-pounds.]

Wind velocity per hour.	Revolutions of wind wheel per minute.	Strokes of pump per minute.	Gallons pumped per minute.	Useful horse-power.
8 miles	18	18	2.8	0.010
12 miles	28	28	4.3	0.014
16 miles	32	32	5.0	0.016
20 miles	35	35	5.4	0.018
25 miles	38	38	5.9	0.019

Mill No. 46.—This is a 10-foot Cornell direct-stroke wooden mill manufactured at Louisville, Kentucky, and is used to pump water for stock. The tower is of wood, 30 feet to the axis of the wheel. The wind wheel has 90 plane sails, each 36 by 4½ by 1½ inches, set at an angle of 47° to the plane of the wheel. The well is a dug well; the water is only about 3 feet below the surface of the ground. The pump has a 3½-inch cylinder and 4 inches stroke. The discharge per stroke was 1 pint when pumping rapidly. The lift was 5⅔ feet. The results of the tests are as follows:

Results of tests of mill No. 46—10-foot wooden Cornell.

[Load per stroke, 6 foot-pounds.]

Wind velocity per hour.	Revolutions of wind wheel per minute.	Strokes of pump per minute.	Gallons pumped per minute.	Useful horse-power.
8 miles	16	16	2.0	0.003
12 miles..............	25	25	3.1	0.004
16 miles..............	36	36	4.5	0.006
20 miles..............	42	42	5.2	0.007
25 miles..............	48	48	6.0	0.007
30 miles..............	52	52	6.5	0.008

Mill No. 47.—This is a 10-foot Dempster steel mill, like that shown in Pl. XII, on a 40-foot steel tower. The wind wheel has 24 curved sails, each 30½ by 13½ by 6½ inches, set at an angle of 29° to the plane of the wheel. The pump is back-geared, 2¼ to 1, and has a 3-inch cylinder and 7 inches stroke. The well is open, 20¼ feet to water, and situated under the porch of a house. The mill is located 76 feet from the well. The cylinder is directly under the mill, in a chamber 4 feet by 4 feet, and 6 feet deep. The lower end of the cylinder was about 12 feet vertically above the surface of the water in the well. The discharge per stroke was 1 quart when pumping rapidly. The lift was about 18 feet. The wind wheel of this mill is not well balanced, and the spring which holds it in the wind is not stiff enough. The results of the tests are as follows:

Results of tests of mill No. 47—10-foot steel Dempster.

[Load per stroke, 37 foot-pounds.]

Wind velocity per hour.	Revolutions of wind wheel per minute.	Strokes of pump per minute.	Gallons pumped per minute.	Useful horse-power.
8 miles...............	37	13	3.2	0.014
12 miles..............	66	23	5.7	0.026
16 miles..............	89	31	7.8	0.035
20 miles..............	95	33	8.3	0.037
25 miles..............	0	0	0	0.000

A diagram which was platted for this mill shows that the curve drops at about 19 miles an hour, and reaches the axis line at 24 miles. Above 24 miles an hour the wheel is entirely out of the wind and does no work. On this diagram was also platted the number of strokes per minute of the 8-foot Dempster (No. 35), to show the effect of the different loads upon the number of strokes per minute. The 8-foot mill was found to be carrying more than three times the load that the 10-foot mill was carrying.

Mill No. 48.—This is a 30-foot Halliday wooden pumping mill on a 70-foot wooden tower. It is owned by the city of Valley Falls, Kansas, and is used to pump water for the city supply. The mill and tower are shown in Pl. XIV. The sail area is arranged in two concentric rings. In the outer ring there are 192 sails, in the inner ring 144 sails, each 43 by 4½ by 3½ inches, set at an angle of 25° to the plane of the wheel. There are two wells; one (11 feet in diameter) directly under the mill, and another (10 feet in diameter) near the bank of the river 375 feet from the mill. A 3-inch suction pipe connects the wells, and a 3-inch supply pipe leads from the lower well to the river. The pump is double-acting, and has a 4-inch cylinder and 11 inches stroke. The water is pumped directly into the distribution pipes, also into an elevated tank. The tank is of wood, 20 feet by 30 feet, and is 5,570 feet distant from the mill. Of the connecting pipe, 50 feet is 3 inches in diameter, 1,200 feet is 4 inches in diameter, and 4,300 feet is 6 inches in diameter. The bottom of the tank is 111 feet above the well platform. The lift or head at any time is, then, the distance from the well to the well platform and 111 feet plus the amount registered on the gage on the tank. The mean lift when the test was made was 135 feet. The cylinder capacity is 2.4 quarts, the measured discharge per double stroke 4.5 quarts. The mean temperature was 90° F., the mean barometric pressure 28.9 inches. The results of the tests are as follows:

Results of tests of mill No. 48—30-foot wooden Halliday.

[Load per stroke, 1,265 foot-pounds.]

Wind velocity per hour.	Revolutions of wind wheel per minute.	Strokes of pump per minute.	Gallons pumped per minute.	Useful horse-power.
8 miles	6	6	6.7	0.23
12 miles	15	15	16.9	0.58
16 miles	20	20	22.5	0.77
20 miles	24	24	27.0	0.92
25 miles	28	28	31.5	1.07

This windmill and pump had been in use about ten years. It furnishes enough water during nine months in the year, but during the months of July and August and part of June and September a steam engine is at times employed to work the pump. The cost of repairs to the mill and pump has been from $50 to $60 a year. The number of strokes per minute for different wind velocities is shown in fig. 27.

This is the largest windmill pumping outfit that we have tested. It is interesting to compare its power with that of the smaller wooden mills and with that of the steel mills.

Mill No. 51.—This is an 8-foot Monitor steel pumping mill on a 30-foot steel tower (see fig. 29). The wind wheel has 18 curved sails, each 31½ by 13 by 5 inches, set at an angle of 35° with the plane of the wheel.   The well is a dug well; depth to water, 18½ feet.   The **water** is used for stock.   The pump has a 3-inch cylinder, 6 inches stroke, **and** 1½-inch suction and discharge pipes.   The cylinder is 1 foot above the lower end of the suction pipe, and is always under water.   A peculiarity of this mill is that the downstroke of the pump

Fig. 29.—Working parts of mill No. 51—8-foot Monitor.

is made in less time than the upstroke.   The mill is back-geared, 35 to 13, 13 of the cogs being passed over on the downstroke and 22 on the upstroke.   This arrangement makes the mill run easily and prolongs its usefulness.   The wind wheel is held in the wind by the weight of the tail; there is no spring.   The lift was 25 feet, the discharge per stroke 0.7 quart.   The mean temperature was 84° F., the mean barometric pressure 27.8 inches. The results of the tests are as follows:

Results of tests of mill No. 51—8-foot steel Monitor.

[Load per stroke, 36.5 foot-pounds.]

Wind velocity per hour.	Revolutions of wind wheel per minute.	Strokes of pump per minute.	Gallons pumped per minute.	Useful horse-power.
8 miles............	27	10	1.7	0.011
12 miles............	56	21	3.7	0.023
16 miles............	81	30	5.2	0.034
20 miles............	91	34	6.0	0.037

When the mill was running uncoupled from the pump, the pump rod made 24 strokes per minute in a 12-mile wind, i. e., the pump load of 36.5 foot-pounds per stroke reduced the number of the strokes of the pump rod from 24 to 21.

For the purpose of ready comparison the principal results of these tests of pumping mills have been tabulated. (See table on p. 64.)

DISCUSSION OF RESULTS OF TESTS.

In this discussion of the results of the tests of these pumping mills we wish to call attention to the principal facts shown by them. The explanation of some of the points is not easy. The useful work which a windmill will do at a given wind velocity depends on several factors, and it is difficult to measure or even estimate the value of each. If the mills could be tested under conditions easily controlled by the experimenter, the problem would be greatly simplified; but each mill is tested under its own conditions of pump, well, wind exposure, and atmosphere. A comparison of the results of the tests of pumping mills with the results of the tests of power mills throws much light on some of the facts (see Part II, pages 107 to 109, inclusive). A fact very evident from the following table is that the useful work done by windmills in pumping water is small. Only one mill, the largest (No. 48), is doing 1 horsepower of useful work in a 25-mile wind. The best 12-foot mill is doing less than 0.64 horsepower and the best 8-foot mill less than 0.12 horsepower in a 25-mile wind.

Results of tests of pumping mills.

No. of mill.	Name of mill.	Size of mill.	Size of pump. b	Lift of pump.	Useful work per stroke.	Number of strokes of pump per minute at given wind velocities (per hour).						Useful horsepower at given wind velocities (per hour).					
		Feet.	Inches.	Feet.	Ft.-lbs.	8 miles.	12 miles.	16 miles.	20 miles.	25 miles.	30 miles.	8 miles.	12 miles.	16 miles.	20 miles.	25 miles.	30 miles.
2	Woodmanse	12	9½x12	17.75	536.2		5.2	16.0	20.3	23.3	25.3		0.085	0.260	0.322	0.379	0.411
3	Aermotor	12	9½x12	13.75	415.3	5.3	12.0	16.4	19.8	23.1	25.0	0.007	0.151	0.207	0.250	0.291	0.315
4	Ideal	8	8x8	12.00	50.0		10.2	19.3	25.3	28.1	25.0		0.015	0.029	0.038	0.043	0.058
5	Aermotor	8	5½x8	13.00	94.9		18.6	25.1	29.8	34.6	38.5		0.053	0.072	0.086	0.099	0.111
6	Gem	8	6½x8	9.00	77.6		12.4	17.9	22.0	25.4	28.5		0.029	0.042	0.051	0.059	0.065
7	Aermotor	12	9½x12	15.50	461.9		11.4	15.8	19.0	22.1	23.5		0.160	0.221	0.266	0.309	0.329
8	Star (a)	10	3x5	30.00	15.0	28.0	30.0					0.013	0.014				
9	Aermotor	16	8x16	44.25	1,013.0		15.0	14.1	17.2	19.6	21.0		0.325	0.453	0.448	0.601	0.644
10	Ideal	8	2½x6	33.00	22.8	14.1	10.6	33.3				0.010	0.017	0.023	0.032		
11	do	12	7½x12	45.00	843.7		25.0	12.7	18.8	23.3	25.0		0.123	0.325	0.481	0.600	0.639
12	do	14	9½x12	11.00	263.5	3.1	4.8	15.1	19.2	21.5		0.025	0.067	0.125	0.153	0.172	
13	Aermotor	12	10x12	11.00	389.0		10.9	17.1	21.0	24.6	27.5		0.110	0.171	0.210	0.247	0.275
15	Gem	10	8x8	15.00	219.0		11.0	11.9	14.7	14.6			0.053	0.082	0.101	0.099	
16	Halliday (a)	12	6x8	16.00	100.0	4.0	7.8	33.9	42.7	52.5		0.012	0.067	0.103	0.130	0.159	
17	Gem	8	8x12	21.75	88.5		22.6	12.8	17.3				0.070	0.149	0.202		
18	Ideal	12	5½x8	14.75	89.2	8.0	6.0	26.1	28.0	27.5		0.022	0.064	0.070	0.076	0.074	
19	Gem	12	10x10	18.00	450.0		20.2	11.9	14.7	16.0			0.085	0.162	0.201	0.219	
20	Jumbo (a)	15½	6x12	14.00	291.2		6.2	5.3	10.7	17.5				0.047	0.086	0.154	
21	Halliday (a)	12	7½x7	15.00	141.5			28.5	37.3	44.6			0.060	0.121	0.159	0.184	
25	Fairbanks	8	2½x4	8.50	5.5		14.0	37.3	38.0	33.3			0.005	0.007	0.008	0.005	
	Irrigator (a)	16		14.00		12.3	29.2		22.0			0.002					
32	Aermotor	12	8x12	30.00	313.0	5.3	14.8	19.0	22.0			0.047	0.14	0.25	0.40	0.44	
33	Woodmanse	10	6x10	22.00	172.0	3.5	14.4	20.5	33.3	24.2		0.020	0.075	0.110	0.195		
35	Dempster	8	3½x8	58.00	133.5		19.2	28.8	30.0	25.7	29.0		0.077	0.116	0.134	0.182	
36	Eclipse (a)	22½	4¼x14	39.00	248.0	4.8	12.2	16.5	20.0	46.0		0.036	0.090	0.124	0.150	0.205	0.223
37	Woodmanse	12	3x9	50.00	254.0	18.0	29.0	19.0	23.3		29.0	0.012	0.094	0.146	0.180	0.029	
38	do. (a)	10	2½x4	27.00	21.0	15.0	30.0	36.0	41.0	46.0		0.017	0.019	0.023	0.026		
39		10	3½x6	22.00	36.0	13.0	22.0	40.0	47.0			0.005	0.034	0.044	0.052		
40	Fairbanks	8	3½x6	28.00	13.0	14.0	25.0	29.0	34.0			0.012	0.009	0.012	0.014		
41	Woodmanse	12	3x6	16.50	29.0	20.0	13.0	33.0	40.0			0.004	0.022	0.029	0.036		
42	Ideal	6	3x4	36.00	55.0	18.0	32.0	19.0	23.0			0.012	0.014	0.020	0.024		
43	Perkins	10	3x4	36.00	57.0	16.0	28.0	41.0	45.0			0.022	0.035	0.045	0.050		
45	Eclipse (a)	10	3x4	55.00	18.0	20.0	25.0	32.0	35.0			0.010	0.014	0.016	0.018		
46	Cornell (a)	10	3½x4½	14.00	6.0	18.0	28.0	36.0	42.0	48.0	52.0	0.003	0.004	0.006	0.007	0.007	0.008
47	Dempster	10	3x7	5.70	6.0	16.0	25.0	31.0	33.0	0	0	0.014	0.026	0.035	0.037	0	0
48	Halliday (a)	30	4x22	18.00	37.0	13.0	23.0	20.0	24.0	28.0		0.23	0.58	0.77	0.92	1.07	
51	Monitor	8	3x6	25.00	36.5	10.0	21.0	30.0	34.0			0.011	0.023	0.034	0.037		

a Wooden mills. b Inside diameter of cylinder by length of stroke.

From the foregoing table it will be seen that mills of the same size differ very much in the amount of useful work, and that some of the larger mills are doing very little more work than some of the smaller ones. Nos. 4 and 18, for example, are the same size and make (the wells, however, are very different), but the latter is doing three or four times more work than the former. The 12-foot mill, No. 11, is doing nearly as much useful work as the 16-foot mill, No. 9, and three or four times more work than the 14-foot mill, No. 12, while it is doing 300 per cent more work than No. 21. The 15½-foot Jumbo will probably do little more work during the season than a good 8-foot mill.

RELATION BETWEEN WIND VELOCITY AND STROKES OF PUMP.

Figs. 6, 10, 11, 12, 13, 15, 17, 18, 19, and 21 show graphically the relation between the wind velocity (in miles per hour) and the strokes of the pump. The curves, as will be noted, differ considerably; but with the exception of fig. 19, for mill No. 20, they agree in that they rise rapidly, reaching the highest point at wind velocities of from 13 to 19 miles. From that point they descend slowly. They differ much in the position of the beginning of the curve, or the velocity required to start the mill. Some will run in an 8-mile wind, while others require a 10-mile or a 12-mile wind to start them. Some rise less rapidly than others, a notable case being mill No. 11. Some descend much more rapidly than others after reaching the highest point. This is especially true of the 8-foot Ideals. Mill No. 20 required a 14-mile wind to start it, and does not appear to have a maximum. The shape of the curve, especially the position of its initial point, is due to the load on the pump, or the number of foot-pounds per stroke. An increase in the load moves the curve to the right and raises it higher. This will be more clearly shown in Part II (see discussion on 12-foot power mill No. 27, pp. 86–89). The height and position of the highest point depend on the tension of the spring, or the weight which holds the mill in the wind. The greater the tension the higher the summit and the farther it is to the right; the less the tension in the spring the steeper the descent from the highest point. The gearing—i. e., the mechanism which causes the pump to make a stroke to each revolution of the wheel, or a stroke every second or third revolution only—modifies the curve. In mills with a direct stroke the curve is much higher and is farther to the right than in back-geared mills, as shown by a comparison of the curves of mills Nos. 3 and 21, shown in figs. 10 and 21.

USEFUL WORK OF PUMPING MILLS.

The relation between wind velocity and horsepower is shown graphically, for five 12-foot mills in fig. 30, and for four 8-foot mills in fig. 31. Examining the five curves of fig. 30, we see that No. 11, the one which gives the greatest horsepower, has the heaviest load and requires the greatest wind velocity to start it. No. 2 has about five-eighths of the load of No. 11, does less work, and requires about the same wind velocity to start it. No. 3 has a lighter load than No. 2, and will start in a wind of about 7 miles an hour. No. 19 has a little heavier load than No. 3, and does much less work at all velocities. The latter requires a 9-mile or a 10-mile wind to start it, while the former will start in a 7-mile or an 8-mile wind. No. 21 is doing the least work of the five, and requires about an 11-mile wind to start it. It is a wooden mill working direct stroke, while the others are steel mills and back-geared.

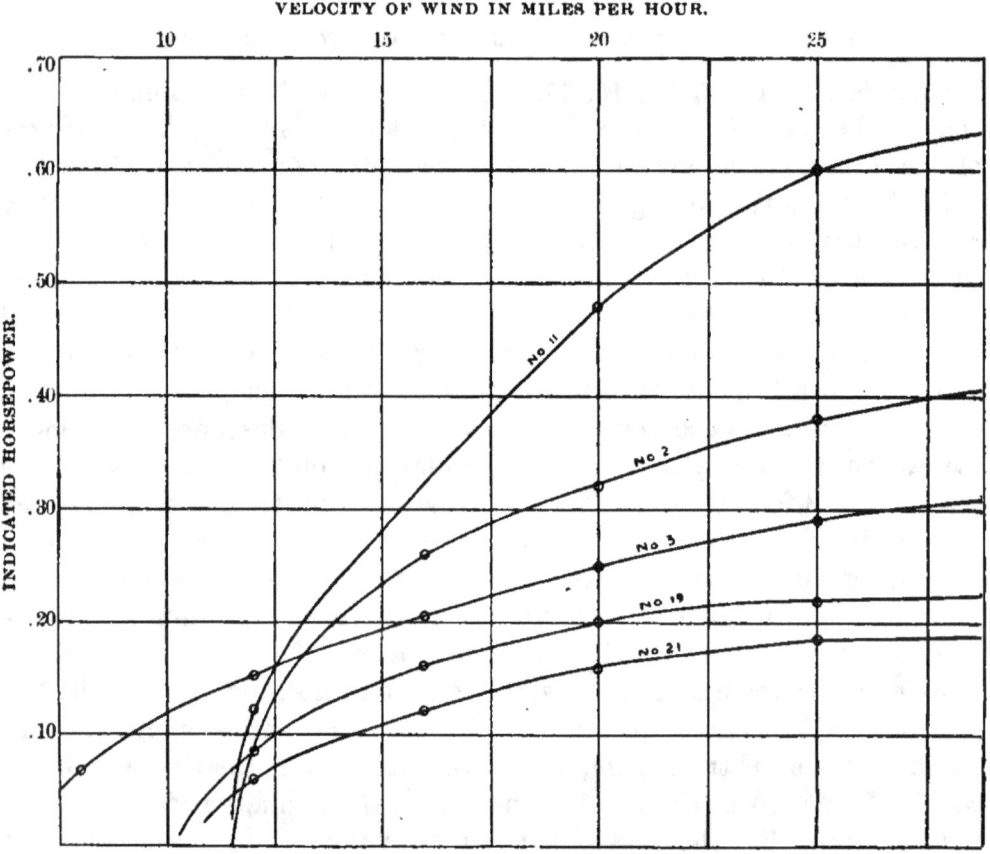

VELOCITY OF WIND IN MILES PER HOUR.

Fig. 30.—Diagram showing relation between horsepower and wind velocity for five 12-foot mills.

It must be understood that in this comparison no correction or allowance is made for the difference in temperature and barometric pressure, nor for the fact that in the case of Nos. 11 and 19 the pumps are on well points, while in the others they are in open wells.

It will be seen that none of the curves in fig. 30 reach a maximum below 30 miles an hour. They do, however, for some higher velocities, since the work per stroke of pump is nearly constant for each pump for all velocities, though not the same for one pump as for another. The curves also give the relation between wind velocity and the number of strokes of pump per minute.

In fig. 31 the curve for No. 18 is seen to reach a maximum at about 25 miles an hour. The others reach their maximum points at velocities of about 30 miles an hour.   These maximum points are points of greatest speed, and are produced by a reduction of wind area, the wind wheel turning out of the wind.   This make of mill is seen to "govern," or turn out of the wind, at a lower velocity than other makes.

Comparing the curves in fig. 31, we see that the pump doing the most work at high wind velocities is No. 5, which is also the one most heavily loaded.   The principal differences between Nos. 4 and 18, the pumps doing the most and the least work for velocities less than 22 miles an hour, are in the load and the well.   No. 4 has five-ninths of the load of No. 18, and is on a well point.   The two pumps doing the least work are on well points.

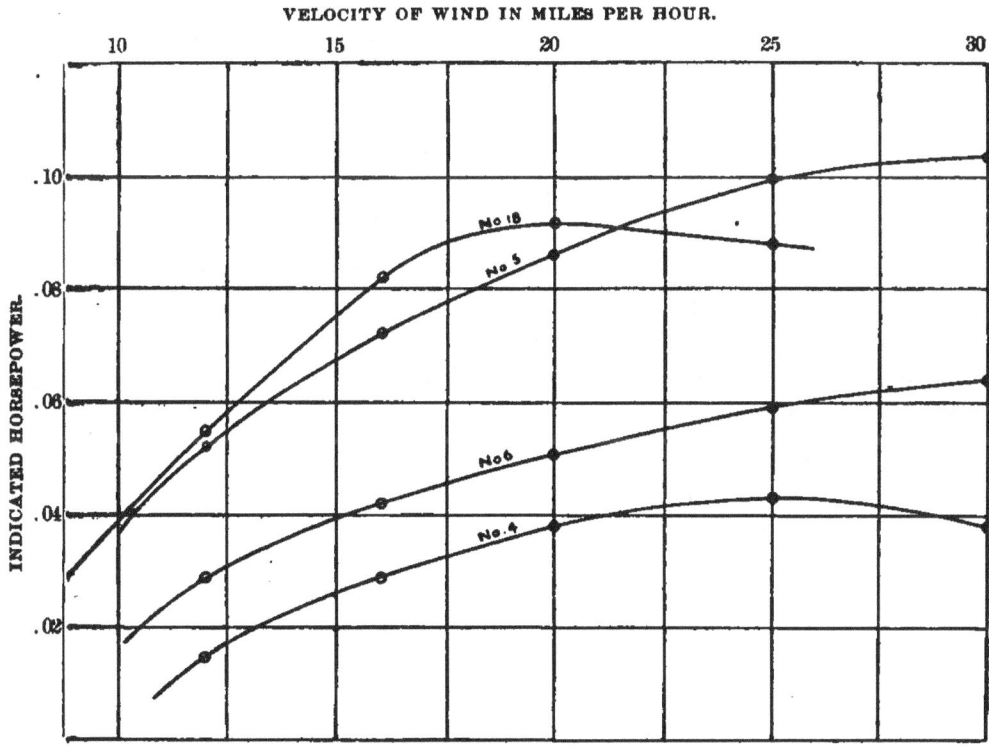

FIG. 31.—Diagram showing relation between horsepower and wind velocity for four 8-foot mills.

Mill No. 25 is used to pump water for stock.   Comparing it with, say, No. 5, we find that its load is about five-ninths as great, and that it is doing about one-tenth as much work as the latter.   It will start with a wind velocity of about 6 miles an hour, while the latter requires a wind velocity of about 8.5 miles an hour.

It is interesting to compare the results of Nos. 35 and 36.   The former is an 8-foot back-geared steel mill, heavily loaded; the latter is a 22.5-foot direct-stroke mill, lightly loaded.   The wind exposure of both is very good.   The discharge per stroke of No. 35 is 1.1 quarts, that of No. 36 about 3 quarts.   The lift of the former is 58 feet, the lift of the latter 39 feet.   The load per stroke of the former is 133 foot-pounds, that of the latter 248 foot-pounds.   No. 35 starts in about a

9-mile wind, No. 36 in a 6-mile wind. In an 8-mile wind the former is doing no work, the latter is making 4.8 strokes per minute; but in a 12-mile wind the former is making 19 strokes per minute, the latter only 12; for all higher velocities the former is making 60 per cent more strokes than the latter. For wind velocities above 12 miles an hour the horsepower of the 22.5-foot mill is only from 7 to 17 per cent greater than that of the 8-foot mill. The difference is really not so great as this, as the actual discharge is not so great as we have supposed.

PRESSURE-TANK SYSTEM.

Mill No. 37 is of special interest, in that the water is forced into a pressure tank instead of into an elevated tank. This system is said to be in use to some extent in the Eastern States, and is just coming into use in the West. The advantage claimed for it is that the tank can be placed in a cellar or under ground, where it will not freeze, instead of on an elevated structure. A hydraulic regulator is used, which causes the mill to turn out of the wind when the pressure in the tank reaches a certain amount. In this case the tank, which is located in a cellar, is 2 feet in diameter and 9.5 feet long; it is about 170 feet from the well and about 43 feet above the surface of the water in the well. The discharge (1.05 quarts per stroke) is pumped into the tank, then forced by the confined air through a 1-inch iron pipe to the hydrants, then through a line of rubber hose to the desired point. The pressure in the tank diminishes with the reduction in the quantity of water in it. For different volumes of air it is as follows:

Pressures in tank for different volumes of air.

Air volume.	Pressure.
Barrels.	*Pounds.*
11.1	0
5.6	14.7
3.7	29.4
2.8	44.1
2.2	58.8
1.9	73.5
1.6	88.2
1.4	103.0

The objection to this system can easily be seen from this table. When the pressure is 103 pounds per square inch the withdrawal of 0.2 barrel of water from the tank lowers the pressure to 88.2 pounds, and the withdrawal of an additional 0.3 barrel lowers the pressure to 73.5 pounds; in other words, the withdrawal of a half barrel of water when the pressure is 103 pounds lowers the pressure 30 pounds. We found, on trial, that when the pump was not working (no wind), and the pressure was at 59 pounds, one hose stream through a $\frac{3}{16}$-inch nozzle reduced the pressure to 50 pounds in one minute and to 45 pounds in two minutes. In seven minutes it fell to 30 pounds, and in seventeen minutes from the time of opening the cock it fell to 20 pounds. In a 20-mile to 30-mile wind, and with one hose stream running, the pump kept the pressure in the tank at about 30 pounds. Where only a small amount of water is needed at a time this system will probably give satisfaction. By the use of a larger tank more water can be obtained for a given reduction of pressure.

It will be noticed that this is the same size and make of mill as No. 2, but that it is altogether different from No. 41. No. 2 is more heavily loaded than No. 37. The heaviest load on No. 37 is 472 foot-pounds per stroke of pump; the heaviest load on No. 2 is 536 foot-pounds per stroke. No. 2 is doing the most useful work, but the hydraulic regulator acts on No. 37 to turn it partly out of the wind for the load 43 plus 75 pounds. No. 41 is doing very little work, but it is lightly loaded and working direct stroke.

COMPARISON OF THREE PUMPING AERMOTORS.

Comparing the 16-foot Aermotor No. 9, the 12-foot Aermotor No. 3, and the 8-foot Aermotor No. 5, a diagram was plotted of the horse-power of these three mills, drawn to the same scale. They were found to start at about the same wind velocity, viz, 7 miles an hour, which indicates about the same total load. In a 25-mile wind the 8-foot mill was found to be yielding 0.1 horsepower, the 12-foot mill 0.29 horsepower, and the 16-foot mill 0.6 horsepower. In other words, the 12-foot mill was doing about three times and the 16-foot mill about six times more work than the 8-foot mill. The conclusion can not be drawn from this that the powers of the mills are to each other in these ratios, since the pump efficiencies are not the same.

USEFUL WORK OF TWO PUMPING MILLS IN A GIVEN TIME.

The useful work which two pumping mills of the same wind area, exposure, pump efficiency, and general character will do depends on the load on the mill and the wind velocity. If the mill is heavily loaded it will do more work at wind velocities of 12 or more miles an hour and less work at lower velocities than one of lighter load. The useful work done in a given time is the product of the work done per hour at the mean velocity multiplied by the number of hours. If the mean velocity at a given place is low, the mill load must be less for maximum work than that at a place where the mean velocity is higher. To illustrate this fact, we will use the results of the tests of two 12-foot pumping mills—No. 11, heavily loaded and giving a greater horsepower at high wind velocities than any other mill tested, and No. 3, giving the greatest power at low velocities. The useful work per stroke of pump is 844 foot-pounds for No. 11 and 415 foot-pounds for No. 3. The useless work of the former is greater than that of the latter, since the pump of the former is on two well points, while the pump of the latter is in an open well. The relation between the horsepower and the wind velocity is shown in fig. 30.   The curves are seen to cross each other at a wind velocity of 12.5 miles an hour.   For less velocities than that No. 3 is doing more work per hour than No. 11, and for greater velocities No. 11 is doing more work than No. 3.   If the velocity were, say, not more than 13 miles an hour, it is very evident that mill No. 3 would do more work in a given time than mill No. 11.

There is no record of wind movement at Garden, Kansas (where most of these tests of pumping mills were made), for any considerable length of time.   There is one, however, for Dodge, 50 miles east of Garden, kept by the United States Weather Bureau, which may be used for this purpose.   The following table gives the number of hours per month for the six months April to September, for the years 1889 to 1895, inclusive, when the wind movement was, respectively, 0 to 5, 6 to 10, 11 to 15, 16 to 20, 21 to 25, 26 to 30, 31 and more miles per hour.

Mean wind movement at Dodge, Kansas, for the seven years 1889 to 1895.

Month.	0–5 miles.	6–10 miles.	11–15 miles.	16–20 miles.	21–25 miles.	26–30 miles.	31 and greater.
	Hours.	*Hours.*	*Hours.*	*Hours.*	*Hours.*	*Hours.*	*Hours.*
April	116	175	157	113	76	43	40
May	116	195	168	120	74	39	32
June	120	187	139	111	86	49	28
July	144	218	176	117	57	23	9
August	178	230	152	99	62	18	5
September ...	166	182	152	93	75	34	18
Mean ...	140	198	157	109	72	34	22

It will be seen from this table that the wind velocity at Dodge is 5 miles or less per hour for 140 hours per month.   During this time neither of these mills (Nos. 3 and 11) will do any work, as neither will start in a 5-mile wind.

The velocity is from 6 to 10 miles an hour for 198 hours a month. Mill No. 3 will start in about a 7-mile wind, and hence will run about four-fifths of this time, or 158 hours.   No. 11 requires 11.5 miles of wind to start it, and will do no work during this time.

The velocity is 11 to 15 miles an hour for 157 hours during the month. No. 3 will work all of this time, and No. 11 about nine-tenths of the time, or 141 hours.

Both mills will run at all higher velocities.   At Dodge, mill No. 3 will run (if in the wind) about 75 per cent of the time, and No. 11 about 51 per cent of the time.   For convenience, these results have been tabulated.

Comparative results of tests of two pumping mills.

Wind velocities, miles per hour.	Mill No. 3.			Mill No. 11.		
	Hours per month.	Horse-power.	Product.	Hours per month.	Horse-power.	Product.
6 to 10	158	0.067	10.6	0	0.00	0.0
11 to 15	157	0.168	26.4	141	0.19	26.8
16 to 20	109	0.230	25.1	109	0.40	43.6
21 to 25	72	0.277	19.9	72	0.56	40.3
26 to 30	34	0.308	10.5	34	0.63	21.4
31 and greater	22	0.320	7.0	22	0.64	14.1
Total	....	....	99.5	....	....	146.2

The second and fifth columns in this table give the number of hours during the mean month that each mill will run with a wind velocity of from 6 to 10 miles an hour. The third and sixth columns give the horsepower for the mean velocity; for example, 0.168 is the horsepower for No. 3 at a wind velocity of 13 miles an hour, and 0.19 is the horsepower for No. 11 at the same velocity. The fourth and seventh columns (the product of the number of hours and the horsepower) give the horsepower that each mill will yield during the month. It will be seen that No. 11 is doing 31 per cent more useful work than No. 3. If this comparison be made for the month of August, it will be found that No. 11 will do 26 per cent more useful work during that month than No. 3.

PROPER LOAD.

It will be seen from what has just preceded, that the useful power of a pumping mill depends to a great extent on its load. If the water is needed constantly and there is little or no storage, as in the case of water for stock, the mill must be lightly loaded, and the useful work it will do is small. If, however, there is plenty of storage, the mill will pump the largest amount of water if heavily loaded. If there were some automatic device for increasing the load on the mill as the wind velocity increases, the problem of proper load would be solved. But such a device seems difficult to construct.

The following table gives data for back-geared steel irrigating mill and good pumps for use in the semiarid regions of the West, especially for Kansas and Nebraska.

Data regarding mills and pumps for use in semiarid regions.

Diameter of mill.	Load per stroke.	Quantity delivered per stroke.	Lift.	Cylinder capacity.	Starting wind velocity (per hour).	Speed per minute in 20-mile wind.	Discharge per hour.	Discharge per 24 hours.
Feet.	*Ft.-lbs.*	*Quarts.*	*Feet.*	*Inches.*	*Miles.*	*Strokes.*	*Gallons.*	*Acre-ft.*
8	125	2.0	30	4.5 × 8	9	27	810	0.059
12	600	9.6	30	8 × 12	8 to 9	20	2,880	0.212
16	1,100	17.6	30	9 × 16	7 to 8	16	4,224	0.311

For half this lift, or 15 feet, the cylinder capacity should be nearly doubled, and for double the lift the cylinder capacity should be about half that given in the table.

In Part II of this paper, published as Water-Supply and Irrigation Paper No. 42, will be found a discussion of the writer's experiments with power mills, a comparison of pumping mills with power mills, a discussion of the effect of tension of spring on the horsepower of mills, a mathematical discussion of the tests of two Aermotors, a discussion of the action of air on the sail of an Aermotor, a discussion of the useful work of two power mills in a given time, discussions of the results of tests of a Jumbo mill and of a Little Giant mill, a comparison of the Little Giant and Jumbo mills, a comparison of the Little Giant mill and the 8-foot Aermotor, a discussion of the indicated and true velocities of windmills, a comparison of the writer's experiments with those of other experimenters, and economic considerations.

[For index see end of Part II, Water-Supply Paper No. 42.]

O

A. VIEW OF MILL NO. 20—15½-FOOT JUMBO.

B. VIEW OF DEFENDER MILL AND PUMP KNOWN AS "WATER ELEVATOR."

VIEW OF MILL NO. 2 (12-FOOT WOODMANSE MOGUL) AND ANEMOMETER.

VIEW OF MILLS NO. 4 (8-FOOT IDEAL) AND NO. 5 (8-FOOT AERMOTOR).

VIEW OF MILL NO. 3—12-FOOT AERMOTOR.

CONTENTS.

ILLUSTRATIONS.

DUTCH WINDMILL AT LAWRENCE, KANSAS.

SCALE

1 2 3 4 5 FT.

ELEVATION OF APPARATUS USED BY PERRY IN WIND-WHEEL TESTS.

THE WINDMILL: ITS EFFICIENCY AND ECONOMIC USE.

PART II.

By EDWARD CHARLES MURPHY.

EXPERIMENTS BY WRITER—CONTINUED.

In Part I of this paper, published as Water-Supply and Irrigation Paper No. 41, will be found a classification of windmills, a discussion of regulating devices, a synopsis of early experiments with windmills, and a discussion of the writer's experiments with pumping mills. This part (II) of the paper contains the results of the writer's experiments with power mills, a comparison of pumping mills with power mills, and discussions of various facts developed by the tests, together with a comparison of the writer's experiments with those of other experimenters, and the economic considerations of the subject.

POWER MILLS.

The power mill differs essentially from the pumping mill in that the latter gives a reciprocating motion to a pump piston, while the former gives a rotary motion to a vertical shaft, and this, in turn, to a horizontal shaft, which drives the grinder or other machine. The mechanism by which this is accomplished in the Aermotor is shown in figs. 33 and 34. The small plane cogwheel makes three revolutions to one revolution of the wind wheel, and the small beveled cogwheel makes two revolutions to one revolution of the small plane wheel; so that the vertical shaft makes six revolutions to one revolution of the wind wheel; or, as we say, the shaft is geared forward 6 to 1. The two beveled cogwheels of the foot gear (fig. 34) change the motion around a vertical axis to a motion around a horizontal axis without changing the rate of speed. In fig. 7, Part I, which shows the pumping mill, it will be seen that the large cogwheel which gives the up-and-down motion to the piston makes one revolution to each 3.3 revolutions of the wind wheel, or that the pump is geared back 3.3 to 1; so that the vertical shaft of a power mill makes twenty revolutions to one stroke of a pump worked by a pumping mill the wind wheel of which is running at the same rate as that of the power mill. The

83

term "geared mill" is sometimes applied to power mills, but inappro-
priately, since the pumping mill also is geared.   The latter is geared
back, the former is geared forward.

Power mills are heavier than pumping mills.   They ordinarily do
more work and carry heavier loads than the latter.   The load on the
pumping mill is constant for all wind velocities, but it may be varied
in the power mill.   The grinder is made so that as the speed increases
the quantity of corn which enters increases, and thus the load and
work done are increased.   The mill is expected to do three or four
kinds of work—for example, pump water, shell and grind corn, and
turn a grindstone.   In a light wind the pump only can be worked,
but as the wind increases one after another of the three other machines
can be set at work, and thus the load be suited to the velocity of the
wind and the mill be made to do the maximum amount of work.   Power
mills are not made smaller than 12 feet in diameter, for the reason
that a small size will not give power enough to be of account except
for pumping.   The ordinary steel power mills are 12 feet, 14 feet, and
16 feet in diameter.

The power that a windmill is capable of developing can be determined better from a power mill than from a pumping mill, because
the efficiency of the pump—which may be anywhere from 20 to 85 per
cent—is eliminated, and because the load on the mill can be varied at
will, and thus the effect of the load on the power of the mill be determined for different wind velocities.

METHOD OF TESTING.

The power was measured by the use of a Prony friction brake placed
on an iron pulley on the foot gear or horizontal shaft.   The brake
was of wood, and had an arm 3 or 4 feet long.   Near the end of this
arm was fastened a spring balance reading to quarters of a pound.
By turning the nuts on the brake the spring balance could be made
to read any desired amount.   As the brake on the pulley was tightened, the reading of the spring balance was increased and the number of revolutions of the shaft decreased.   The brake is shown in Pl.
XV.   The speed of the shaft was found in one of three ways—which-
ever was most convenient.   A small electric device was used whenever it could conveniently be attached to the wind wheel.   The clicks
of this instrument could easily be counted, and gave the number of
revolutions of the wind wheel for each half mile of wind movement.
A speed counter was used, but did not prove satisfactory.   Whenever
the electric device could not conveniently be employed, the number
of revolutions of the wind wheel was found by counting the revolutions of a mark on the wind wheel as reflected in a mirror conveniently placed.

To illustrate: If u is the number of revolutions per minute of the
brake pulley as found from the revolutions of the wind wheel per

half mile of wind, L the load in pounds as read from the spring balance, and R the length of the arm, then the useful work, in foot-pounds per minute, is—$W = 2 \pi R u L$, and the horsepower is—H. P. $= 2 \pi R u L \div 33,000$.

The number of revolutions per minute of wind wheel was found for each mill for from two to six different brake loads, for wind velocities as small as would keep the mill working for the particular load used

FIG. 32.—Working parts of mill No. 26—14-foot Perkins.

up to about 25 miles an hour. That is, the reading of the spring balance was kept as nearly constant as possible until we had obtained points and a curve like that shown in fig. 10. Then the load was changed and the tests continued in the same way, getting another curve. From these curves the number of revolutions of wind wheel per minute for different loads and wind velocities was easily found, as before indicated. These are given in the table of results for each mill tested, and in many cases are also shown by diagram. The horsepower of any mill for different loads and velocities is easily found by the foregoing formula. These are given in the tables of results of tests, and are also shown by diagrams.

MILLS TESTED.

Mill No. 26.—This is a 14-foot Perkins steel power mill on a 40-foot steel tower, made by the Perkins Windmill Company, of Mishawaka, Indiana. The working parts are shown in fig. 32. The wind wheel has 32 curved sails, each 41 by 14.25 by 7.75 inches, set at an angle of 31° with the plane of the wheel. The shaft is geared forward 6 to 1. The radius of the brake pulley was 5 inches, the length of brake arm 33.5 inches. This mill was tested twice. Between the dates of testing some repairs were made to the shafting, causing the cogwheels to bind less tightly. The following figures are those obtained from the second test. The mill had been in use only about one year, and showed very poor workmanship. The results of the test are as follows:

Results of test of mill No. 26—14-foot steel Perkins.

Load on brake.	Load per revolution of wind wheel.	Number of revolutions of wind wheel per minute at given wind velocities (per hour).					Useful horsepower at given wind velocities (per hour).				
		8 miles.	12 miles.	16 miles.	20 miles.	25 miles.	8 miles.	12 miles.	16 miles.	20 miles.	25 miles.
Pounds. 6	*Ft.-lbs.* 645	......	......	16	31	48	......	......	0.313	0.609	0.937

Mill No. 27.—This is a 12-foot Aermotor on a 30-foot steel tower. The wind wheel is like that of mill No. 3 (see pp. 29 to 30, Part I). The horizontal shaft is geared forward 6 to 1. Fig. 33 shows the working parts, and fig. 34 the foot gear. The brake pully was 9.5 inches in diameter and was fastened to the foot gear at *a* in fig. 34. The brake arm was 35.25 inches in length. The mean temperature during the test was 46° F., and the mean barometric pressure 28.9 inches. The results of the tests are as follows:

Results of tests of mill No. 27—12 foot Aermotor.

Load on brake.	Load per revolution of wind wheel.	Number of revolutions of wind wheel per minute at given wind velocities (per hour).					Useful horsepower at given wind velocities (per hour).				
		8 miles.	12 miles.	16 miles.	20 miles.	25 miles.	8 miles.	12 miles.	16 miles.	20 miles.	25 miles.
Pounds. 0	*Ft.-lbs.* 0	30	49	63	75	87	0.089	0.285	0.386	0.458	0.523
2	222	16	43	57	70	81		0.303	0.653	0.890	1.02
4	444	......	23	48	65	77				1.03	1.45
6	666			12	50	72					

The revolutions of wind wheel per minute for the four brake loads 0, 2, 4, and 6 pounds, respectively, are shown in fig. 35. The pull necessary to overcome the frictional resistance was found by standing on the platform of the mill and slowly turning the wind wheel around with a spring balance. This was checked by winding a

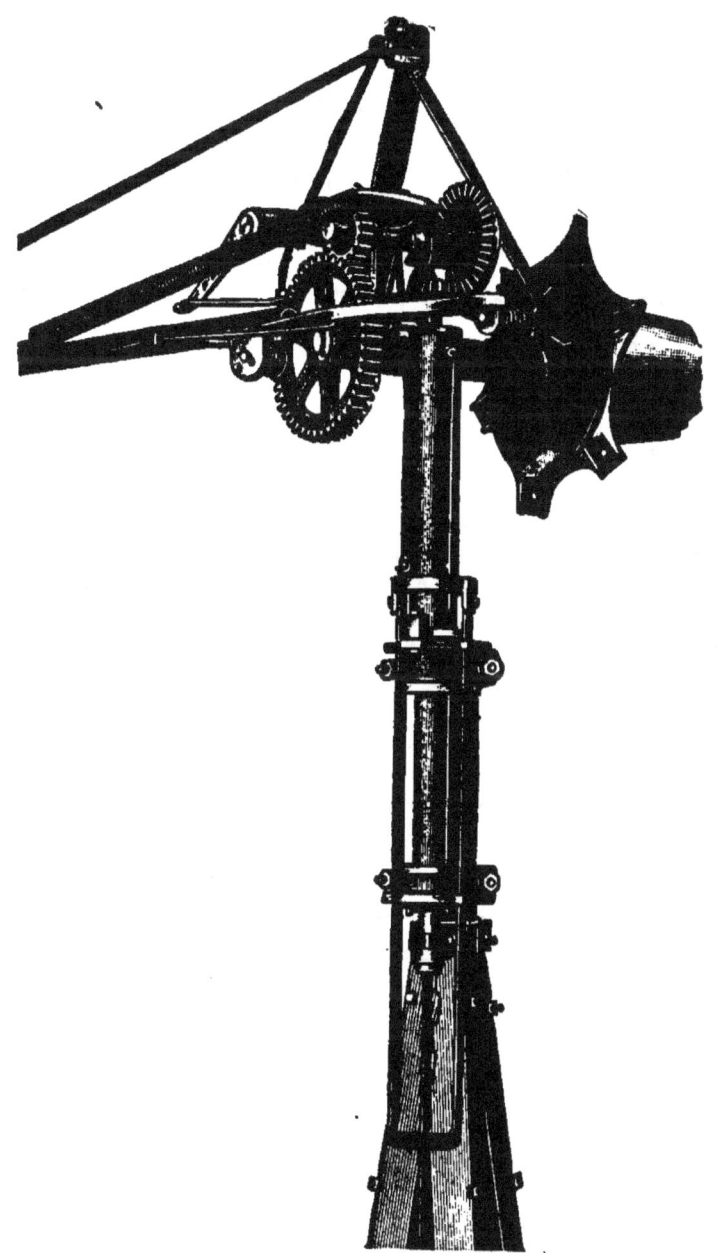

FIG. 33.—Working parts of mill No. 27—12-foot Aermotor.

cord around the circumference of the wind wheel, and, standing on the ground, moving the wheel when there was no wind by pulling on the spring balance attached to the cord. A pull of 1.25 pounds applied at the circumference was sufficient to overcome this resistance at a low velocity. The work done in overcoming this resistance is $1.25 \times 2\pi \times 6 = 47.1$ foot-pounds per revolution. The work done per revolution of wind wheel per pound on the brake arm is—$2\pi \times 35.25 \times 6 \div 12 = 111$ foot-pounds. The ratio of these is $47.1 \div 111 = 0.425$ pound. Hence a brake load of 0.425 pound is equivalent to the friction load.

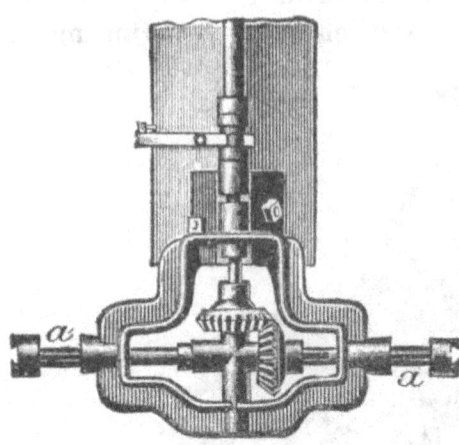

FIG. 34.—Foot gear of mill No. 27—12-foot Aermotor. *a* indicates point where brake pulley was attached.

The effect of each additional 2 pounds load on the brake in reducing the speed of the wheel at different wind velocities is clearly shown here. It is seen that an added load makes a greater proportionate reduction in the speed when the velocity is low than when it is high. Thus, in an 8-mile wind the addition of 2 pounds to the load reduces the speed 50 per cent, while in a 25-mile wind the same load reduces the speed only about 7 per cent.

It will be seen that for wind velocities above a certain amount, with the load not too great, each additional pound of load reduces the speed of the wheel by about the same amount. For example, in a 25-mile wind the addition of 2 pounds changes the speed from 87 revolutions to 81 revolutions. The addition of 2 pounds more changes

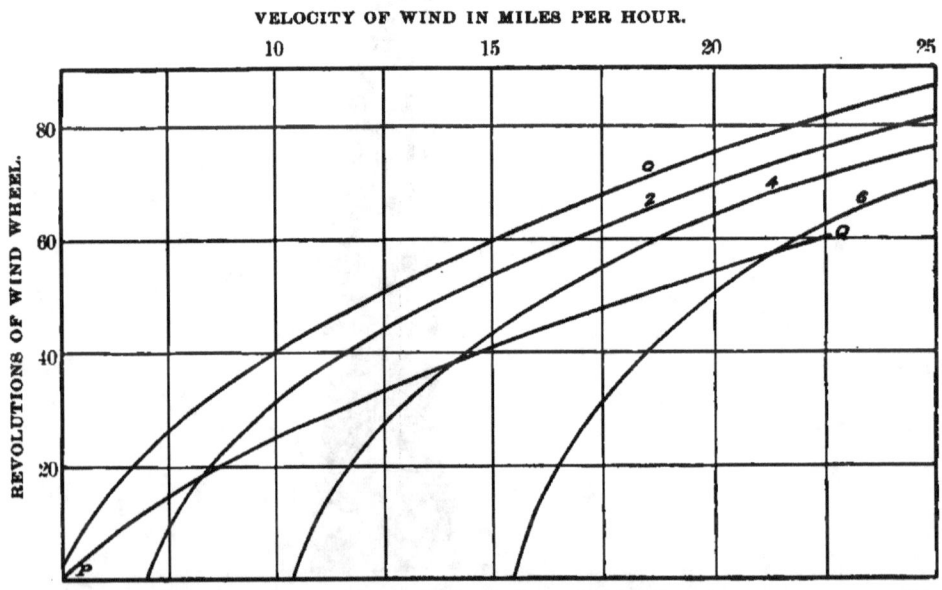

FIG. 35.—Diagram showing revolutions of wind wheel of mill No. 27—12-foot Aermotor. Curve marked *o* is for no brake load; curve marked 2 is for a brake load of 2 pounds; curve marked 4 is for a brake load of 4 pounds; curve marked 6 is for a brake load of 6 pounds; curve *PQ* is speed curve for maximum power.

the speed from 81 revolutions to 77 revolutions. It will be seen that as the load increases the increment of wind velocity necessary to start the mill increases more rapidly than the increment of loading. That is, the 0-pound load curve starts in about a 4.5-mile wind, the 2-pound curve in about a 7-mile wind, the 4-pound curve in about a 10.5-mile wind, and the 6-pound curve in about a 15.5-mile wind. The difference between these starting velocities is constantly increasing. A diagram was platted showing the horsepower of this mill for the 2-pound, 4-pound, and 6-pound brake loads. The curves showed that for any brake load the power of the mill increased rapidly as the wind velocity increased, and that it reached a maximum for some velocity greater than 30 miles an hour. As the load increased the velocity required to start the mill increased rapidly and the curve became steeper. For a given wind velocity the power increased rapidly as the load increased. For a velocity of 25 miles an hour the power was nearly proportional to the load for loads of less than 6 pounds. It showed that when the velocity was less than 12 miles an hour a 4-pound load was too great, and when it was less than 19 miles an hour a 6-pound load was too great. It showed also that the efficiency decreased as the wind velocity for a given load increased, and that it increased as the load increased. The efficiency for a load of 2 pounds and a wind velocity of 9 miles an hour was 40 per cent. At 14 miles an hour and with a 4-pound load it was 36 per cent. If the load at that velocity was reduced to 2 pounds, the efficiency was reduced to 24 per cent. Finding the efficiency by using the wind area (area of circle 12 feet in diameter) instead of the sail area, as is sometimes done, the foregoing efficiency of 40 per cent with a 2-pound load in a 9-mile wind became 26 per cent.

The results for this mill will be discussed from a mathematical point of view further on.

Mill No. 28.—This is a 16-foot Althouse wooden power mill manufactured by Althouse, Wheeler & Company, of Waupun, Wisconsin. It is shown in fig. 36. The axis of the wind wheel is 32 feet above the ground and 15 feet above the roof of a near-by blacksmith's shop. The wind wheel has 130 sails, each 48 by 4 by 1.5 inches, set at an angle of 32° to the plane of the wheel. Two half sails are missing and two others are slightly injured, making a loss of about one and a half sails. The horizontal shaft, which works a sheller, grinder, emery wheel, and wood saw, is geared forward 8.377 to 1. A pull of from 7 to 13 pounds at a distance of 6 feet from the center was necessary to start the mill, showing it to be a hard-running one. The brake pulley was 8 inches in diameter, the brake arm 3.5 feet long. A second visit to this mill was necessary in order to get results for high velocities. During the interval between the tests a 510-pound fly wheel was put on the shaft, which steadied the motion of the mill somewhat. The owner is well pleased with the action of this balance wheel. Single measurements of the power of this mill for the same load and wind velocity differ considerably. It is very evident that this mill is not high enough; if it were 30 or 40 feet higher it would give better results.   It is with the aid of better results obtained from tests of other mills of similar make that we are able to give the results in the following table and in the diagrams.

FIG. 36.—View of mill No. 28—16-foot wooden Althouse.

Results of tests of mill No. 28—16-foot wooden Althouse.

Load on brake.	Load per revolution of wind wheel.	Number of revolutions of wind wheel per minute at given wind velocities (per hour).					Horsepower of mill at given wind velocities (per hour).				
		8 miles.	12 miles.	16 miles.	20 miles.	25 miles.	8 miles.	12 miles.	16 miles.	20 miles.	25 miles.
Pounds.	*Ft.-lbs.*										
0	0	13	23	30	36	40					
1¾	427		20	27	31	36		0.26	0.35	0.40	0.46
G.			5	13	20	28					
5¾	914			19	23	27			0.52	0.64	0.75
8¾	1,462			8	13	15			0.35	0.60	0.67

Fig. 37 shows the number of revolutions per minute of the wind wheel of this mill for the brake loads 0, 1.75, 5.75, and 8.75 pounds, and for grinder load. The curve for the grinder load is a nearly straight line. This is due to the fact that the grinder is constructed

so that as its speed increases the amount of corn it receives increases; thus the load increases automatically as the wind velocity increases. By comparing these speed curves, as they may be called, with those of fig. 35, for the Aermotor, it will be seen that the speed of the latter is much greater than that of mill No. 28.

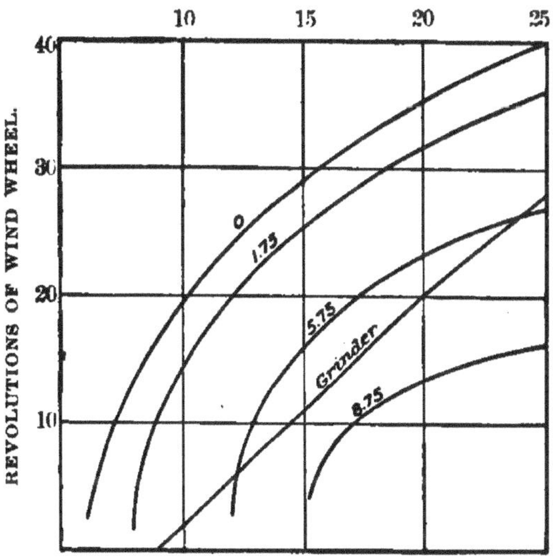

Fig. 38 shows the horsepower of this mill for three brake loads— 1.75 pounds, 5.75 pounds, and 8.75 pounds. The latter load is too great for the mill. By comparing the results for this mill with those for the 12-foot Aermotor (fig. 35) it will be seen

FIG. 37.—Diagram showing revolutions of wind wheel of mill No. 28—16-foot wooden Althouse. Curves marked 0, 1.75, 5.75, 8.75, and grinder are for brake loads of 0, 1.75, 5.75, and 8.75 pounds, respectively, and for grinder load.

that the latter mill is superior to the 16-foot wooden mill.

Mill No. 29.—This is a 16-foot Aermotor like that shown in Pl. XV.

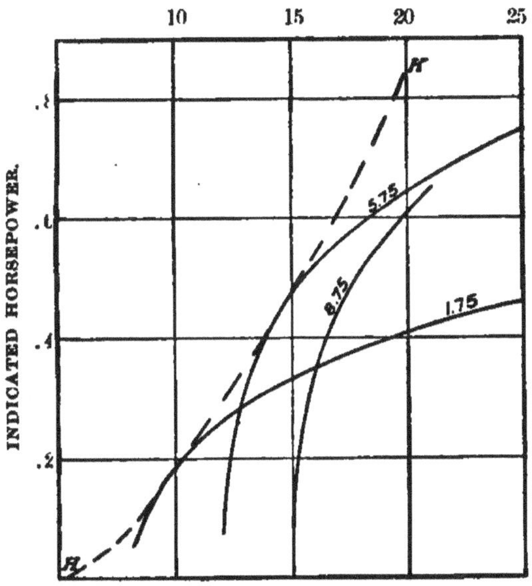

FIG. 38.—Diagram showing horsepower of mill No. 28—16-foot wooden Althouse. Curves marked 1.75, 5.75, and 8.75 show horsepower for brake loads of 1.75, 5.75, and 8.75 pounds, respectively; dotted curve HK shows maximum power.

It is manufactured by the Aermotor Company, of Chicago, Illinois. The tower is of wood, 42 feet to the axis of the wheel. The wind wheel has 18 curved sails, each 59 by 25.75 by 10.5 inches, set at an angle of 30° to the plane of the wheel. It is used for shelling and grinding corn and for working a small pump. The shafting is 20 feet above the ground, near the roof of a granary. It is geared forward 6 to 1, and arranged so that the pump makes 1.011 strokes to each revolution of the wind wheel. The pump lifts 0.022 gallon per stroke a distance of about 40 feet. It has a cylinder 1.5 inches in diameter and a stroke of 8 inches. The supply pipe is on a well point.

The brake pulley is 12 inches in diameter, the brake arm 3.75 feet long. The test was continued until the shafting failed. The mean barometric pressure was 27.8 inches, the mean temperature 70° F.   This mill had been in nearly constant use about five years.   It replaced a 22.5-foot Halliday.   The owner claims that the 16-foot steel mill does more work than the 22.5-foot Halliday wooden mill did.   The results of the tests are as follows:

Results of tests of mill No. 29—16-foot Aermotor.

Load on brake.	Load per revolution of wind wheel.	Number of revolutions of wind wheel per minute at given wind velocities (per hour).					Horsepower of mill at given wind velocities (per hour).				
		8 miles.	12 miles.	16 miles.	20 miles.	25 miles.	8 miles.	12 miles.	16 miles.	20 miles.	25 miles.
Pounds	*Ft.-lbs.*										
1¼	212	20	35	......	......	......	0.13	0.23	......	......	......
P+1¼	487	11	30	41	......	......	0.17	0.44	0.636	......	......
P+2¼	629	......	25	35	......	......		0.48	0.714	......	......

Fig. 39 shows the number of revolutions of the wind wheel for three loads—212 foot-pounds, 487 foot-pounds, and 629 foot-pounds. Although these results are incomplete, on account of the failure of the shafting, they are complete up to a wind velocity of 15 miles an hour, and when studied in connection with the complete test of a mill of the same size and make (No. 44) it will be seen that this mill has about the same power for the same loads at any given wind velocity.

Mill No. 30.—This is a 16-foot wooden power mill known as an Irrigator, used for lifting water. (For description see pp. 49–50, Part I.) Two brake loads (2 pounds and 16 pounds) were used on an arm 2.5 feet long. Curves showing the number of revolutions of the wind wheel per minute for these loads and the useful elevator load are reproduced in fig. 22, Part I; the horsepowers for these loads are shown in fig. 23, Part I.

Mill No. 31.—This is a 14-foot Elgin wooden power mill used to lift water with a rotary (Wonder) pump. It is described on pages 50 to 51, Part I.

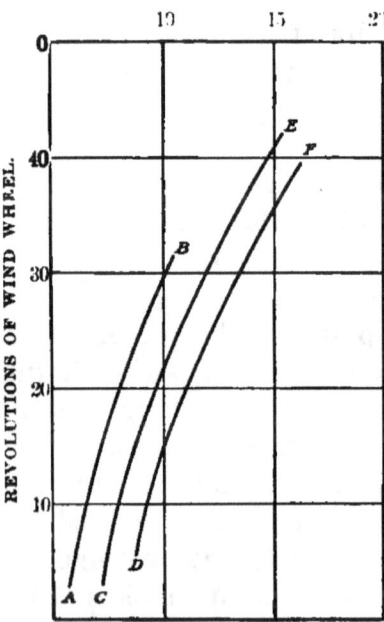

VELOCITY OF WIND IN MILES PER HOUR.

FIG. 39.—Diagram showing revolutions of wind wheel of mill No. 29—16-foot Aermotor. Curve *AB* is for a load of 212 foot-pounds; *CE* is for a load of 487 foot-pounds; *DF* is for a load of 629 foot-pounds.

Mill No. 34.—This is a 14-foot Junior Ideal steel power mill manufactured by the Stover Manufacturing Company, of Freeport, Illinois. The tower is of wood, 41 feet to the axis of the wheel.   The wheel has

24 curved sails in eight sections, regulated on the centrifugal principle. Each sail is 49 by 18 by 8 inches, set at an angle of 29° to the plane of the wheel. This is a sectional vaneless mill. In place of a vane there is a counterpoise. It is geared forward 8 to 1. The mill is used for shelling and grinding corn and elevating. The brake pulley is on a line shaft 15 or 20 feet long. In a 12-mile wind the mill ground 12 pounds quite fine for a mile of wind, or at the rate of 144 pounds an hour. In an 18-mile wind it ground 26 pounds for a mile of wind, or at the rate of 468 pounds an hour. The grinder was made by the Baker Manufacturing Company, of Evansville, Wisconsin. The wind was unsteady, the temperature high—100° in the shade at noon. The results of the tests are as follows:

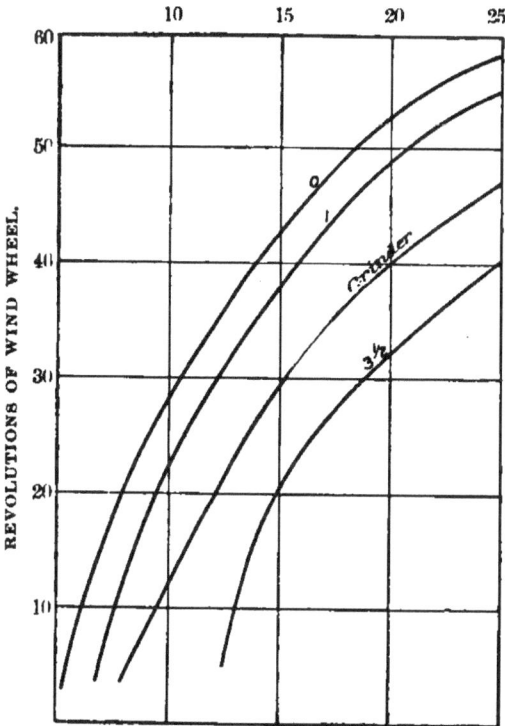

FIG. 40.—Diagram showing revolutions of wind wheel of mill No. 34—14-foot Junior Ideal. Curves marked 0, 1, 3¼, and grinder are for brake loads of 0, 1, and 3.5 pounds, respectively, and for grinder load.

Results of tests of mill No. 34—14-foot steel Junior Ideal.

Load on brake.	Load per revolution of wind wheel.	Number of revolutions of wind wheel per minute at given wind velocities (per hour).					Horsepower of mill at given wind velocities (per hour).				
		8 miles.	12 miles.	18 miles.	20 miles.	25 miles.	8 miles.	12 miles.	18 miles.	20 miles.	25 miles.
Pounds.	*Ft.-lbs.*										
0	0	20	34	44	53	58	.....	.....	.....	.....	.....
1	180	13	29	40	49	55	0.07	0.16	0.22	0.27	0.33
G.	G.	4	19	32	39	47	.....	.....	.....	.....	.....
3¼	631	0	0	24	32	40	0	0	0.46	0.61	0.76

Fig. 40 shows the number of revolutions of the wind wheel per minute for brake loads of 0, 1, and 3.5 pounds, and for the grinder load. Fig. 41 shows the horsepower for brake loads of 1 and 3.5 pounds.

Mill No. 44.—This is a 16-foot Aermotor on a 40-foot steel tower. (See Pl. XV.) The working parts of the mill are like those shown in fig. 33; the foot gear is like that shown in fig. 34. The sail area is the same as that of mill No. 29, page 91. The power was measured with a wooden brake having an arm 4.67 feet long, on a 10-inch iron pulley on the foot gear. Five brake loads were used—0, 3, 5, 8, and 11 pounds, respectively. The shafting is geared forward 6 to 1.

The results of the tests are as follows:

Results of tests of mill No. 44—16-foot Aermotor.

Load on brake.	Load per revolution of wind wheel.	Number of revolutions of wind wheel per minute at given wind velocities (per hour).					Horsepower of mill at given wind velocities (per hour).				
		8 miles.	12 miles.	16 miles.	20 miles.	25 miles.	8 miles.	12 miles.	16 miles.	20 miles.	25 miles.
Pounds.	*Ft.-lbs.*										
0	0	23	38	48.0	56	64.5					
3	528	......	28	41.0	50	58.5	......	0.45	0.66	0.80	0.94
5	880	......	13	33.5	44	53.5	......	0.35	0.89	1.16	1.43
8	1,408	......	......	16.0	36	47.0	......	......	0.68	1.53	2.01
11	1,936	......	......		25	39.5	......	......		1.47	2.31

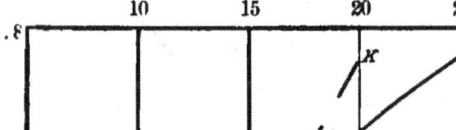

VELOCITY OF WIND IN MILES PER HOUR.

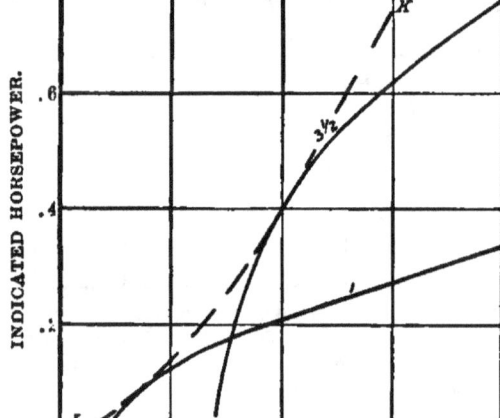

FIG. 41.—Diagram showing horsepower of mill No. 34—14-foot Junior Ideal. Curves marked 3¼ and 1 show power for brake loads of 3.5 pounds and 1 pound, respectively; dotted curve *IK* shows maximum power.

Fig. 42 shows the number of revolutions per minute of the wind wheel for these brake loads. The number on each curve indicates the brake load for that curve. These curves are seen to closely resemble the corresponding curves for the 12-foot Aermotor (fig. 35). The 16-foot mill will be seen to start, with no load, in about a 4.5-mile wind—the same as the 12-foot Aermotor. Fig. 43 shows the horsepower of this mill for loads of 3, 5, 8, and 11 pounds, respectively. The curves for this mill closely resemble those of the 12-foot Aermotor. The curves of the latter were platted, but the diagram is not reproduced because of lack of space. It will be shown further on that these load curves are parabolas, and hence that the power increases as the square root of the wind velocity. It will be shown also that the curve of maximum power is a parabola, that the load for it increases nearly as the first power of the wind velocity, and that the speed of the wheel increases also as the first power of the wind velocity.

Mill No. 49.—This is a

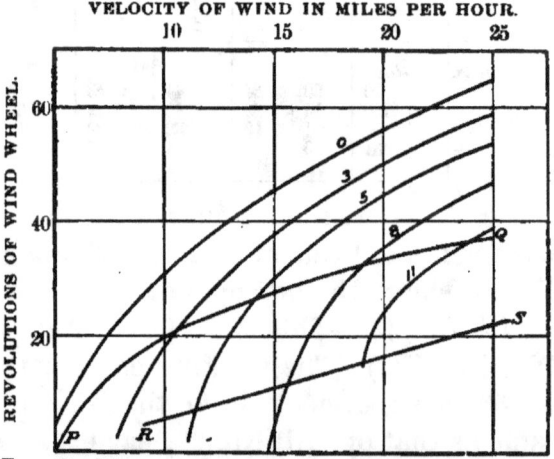

FIG. 42.—Diagram showing revolutions of wind wheel of mill No. 44—16-foot Aermotor. Curves marked 0, 3, 5, 8, and 11 are for brake loads of 0, 3 pounds, 5 pounds, 8 pounds, and 11 pounds, respectively; curve *PQ* is speed of wheel for maximum load; *RS* is load curve for maximum power.

22.5-foot Halliday wooden power mill on a 43-foot wooden tower. The sail area is in two concentric rings, the outer ring having 144 sails, the inner ring 100 sails, each 43 by 4.5 by 3.5 inches, set at an angle of 25° to the plane of the wheel. The upper gearing has a ratio of 50 to 14, the lower gearing a ratio of 53 to 26, so that the horizontal shaft is geared forward 7.28 to 1. The brake pulley is 8 inches in diameter, the brake arm 4.75 feet long. The mean temperature was 82° F., the mean barometric pressure 28.7 inches. The mill is used for shelling and grinding corn. Four brake loads (0, 1.5, 5, and 9 pounds, respectively) were used, also the grinder load.

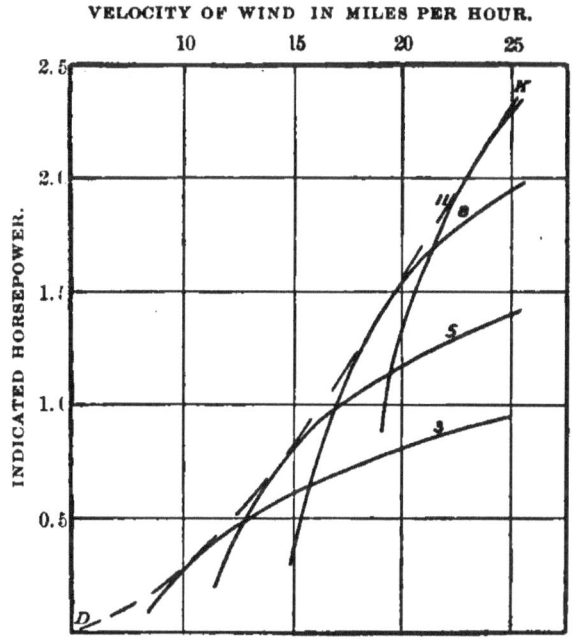

FIG. 43.—Diagram showing horsepower of mill No. 44—16-foot Aermotor. Curves marked 3, 5, 8, and 11 show power for brake loads of 3, 5, 8, and 11 pounds, respectively; dotted curve DK shows maximum power.

The results of the tests are as follows:

Results of tests of mill No. 49—22.5-foot wooden Halliday.

Load on brake.	Load per revolution of wind wheel.	Number of revolutions of wind wheel per minute at given wind velocities (per hour).					Horsepower at given wind velocities (per hour).				
		8 miles.	12 miles.	16 miles.	20 miles.	25 miles.	8 miles.	12 miles.	16 miles.	20 miles.	25 miles.
Pounds.	*Ft.-lbs.*										
0	0	11	20	25	30	32					
1.5	326	6	16	22	26	30	0.059	0.163	0.217	0.257	0.296
5.0	1,087	......	11	18	22	26	......	0.342	0.593	0.724	0.856
G.	......	......	8	16	22	......					
9.0	1,956	......	2	10	15	19	......	0.118	0.593	0.890	1.126

In a 13.5-mile wind the mill ground 20 pounds of chop quite fine in 4.5 minutes. Fig. 44 shows the number of revolutions per minute of the wind wheel for the five loads. This mill requires a 5.5-mile wind to start it without any load, and it makes only 32 revolutions in a 25-mile wind. This is about half as many as are made under the same conditions by the 16-foot mill No. 44. Since the circumference of the 22.5-foot mill is 1.4 times greater than that of the 16-foot mill, the circumference velocity of the 16-foot mill is 44 per cent greater than that of the 22.5-foot mill. The dotted curve, showing the speed of the wheel for the grinder load, will be seen to be a nearly straight line, showing a constantly increasing load with increase of wind velocity. Fig. 45 shows the horsepower for three loads, also the maximum horsepower. It will be seen that the power is small for so large a mill. The shafting of this mill is very heavy, and the grinder is run by a belt from the main shaft. The mill, although on a 43-foot tower, should be at least 20 feet higher. It will be seen to be a very poor mill.

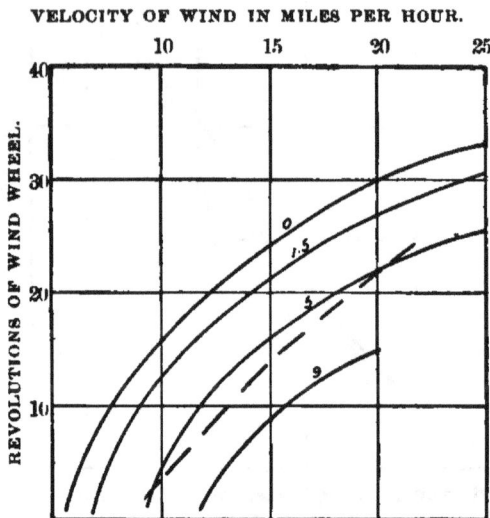

VELOCITY OF WIND IN MILES PER HOUR.

FIG. 44.—Diagram showing revolutions of wind wheel of mill No. 49—22.5-foot wooden Halliday. Curves marked 0, 1.5, 5, and 9 are for brake loads of 0, 1.5, 5, and 9 pounds, respectively; the dotted curve shows the speed of wheel for grinder load.

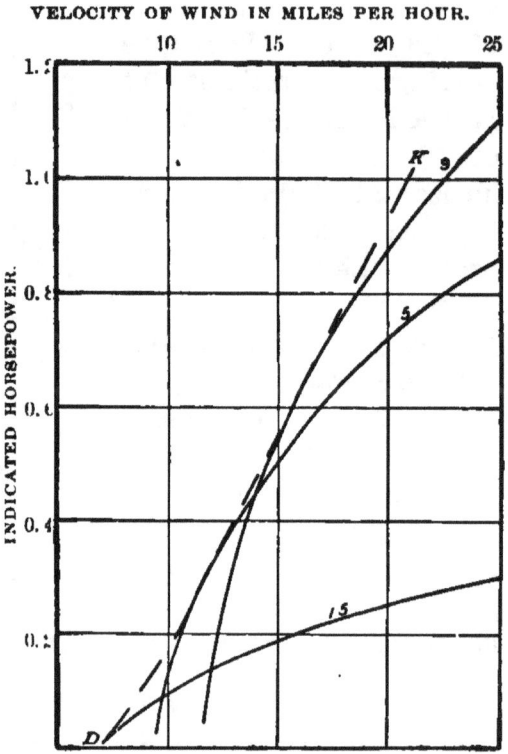

VELOCITY OF WIND IN MILES PER HOUR.

FIG. 45.—Diagram showing horsepower of mill No. 49—22.5-foot Halliday. Curves marked 9, 5, and 1.5 pounds show horsepower for brake loads of 9, 5, and 1.5 pounds, respectively; the dotted line DK shows maximum horsepower.

Mill No. 50.—This is a 12-foot Monitor wooden power mill on a 36-foot wooden tower. (See fig. 46.) It is a sectional mill and has 96 sails, each 44 by 4.25 by 1.75 inches, set at an angle of 34° to the plane of the wheel. The shaft is geared forward 3.66 to 1. The swivel gearing, which enables the mill to turn easily and keep full in the wind, is shown in fig. 47. The mill is used for shelling and grinding corn and pumping water. It is in very good condition, and the wind exposure is very good. The mill had been in use about three years. The mean temperature during the time of test was 83° F., the mean barometric pressure 28.6 inches.

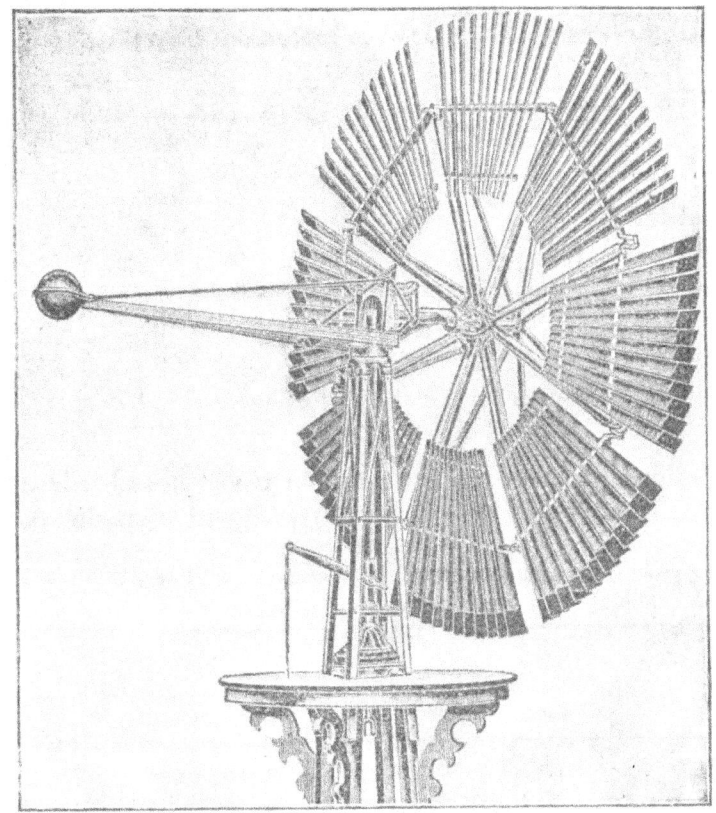

FIG. 46.—Mill No. 50—12-foot wooden Monitor.

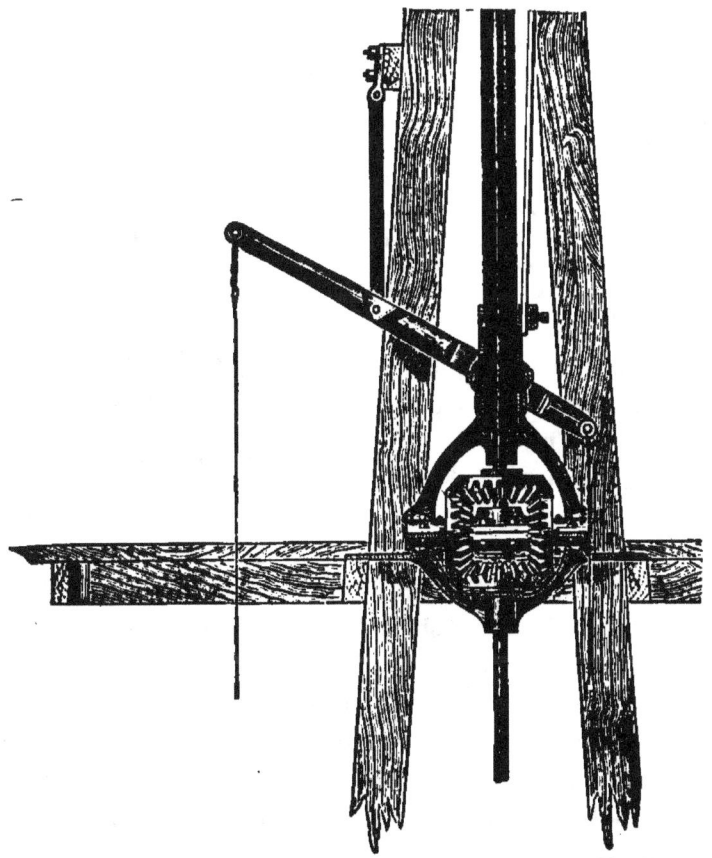

FIG. 47.—Swivel gearing of mill No. 50—12-foot wooden Monitor.

The results of the tests are as follows:

Results of tests of mill No. 50—12-foot wooden Monitor.

Load on brake.	Load per revolution of wind wheel.	Number of revolutions of wind wheel per minute at given wind velocities (per hour).					Horsepower of mill at given wind velocities (per hour).				
		8 miles.	12 miles.	16 miles.	20 miles.	25 miles.	8 miles.	12 miles.	16 miles.	20 miles.	25 miles.
Lbs.	*Ft.-lbs.*										
0	0	16	33	44	54	64	......	......	......	......	......
1.5	120	......	25	35	43	52	......	0.091	0.127	0.156	0.189
2⅞	314	......	......	27	34	40	......	......	0.230	0.324	0.381
4.5	490	......	......	10	24	30	......	......	0.150	0.357	0.445

The revolutions of the wind wheel for three brake loads (0, 1.5, and 2.875 pounds) are shown in fig. 48. This figure also shows the num-

VELOCITY OF WIND IN MILES PER HOUR.

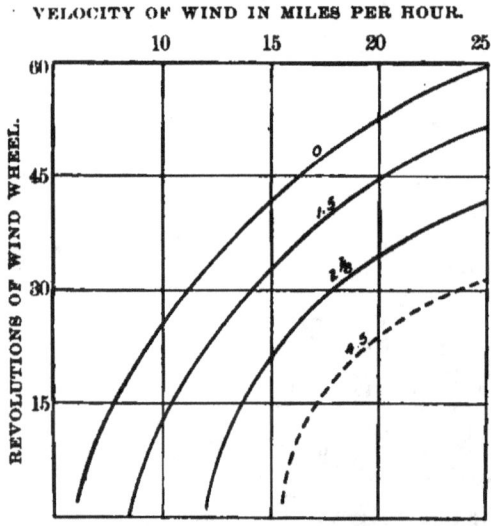

FIG. 48.—Diagram showing revolutions of wind wheel of mill No. 50—12-foot wooden Monitor. Curves marked 0, 1.5, 2⅞, 4.5 are for brake loads of 0, 1.5, 2⅞, and 4.5 pounds, respectively.

VELOCITY OF WIND IN MILES PER HOUR.

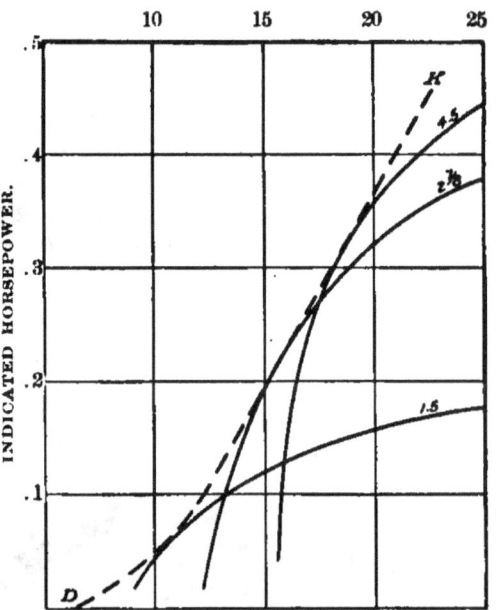

FIG. 49.—Diagram showing horsepower of mill No. 50—12-foot wooden Monitor. Curves marked 4.5, 2⅞, and 1.5 are for brake loads of 4.5, 2⅞, and 1.5 pounds, respectively; dotted curve DK shows maximum power.

ber of revolutions for a 4.5-pound load, found by interpolation from the other results. This mill will be seen to require about a 6-mile wind to start it without any load and to make only 64 revolutions per minute in a 25-mile wind. The 12-foot Aermotor will start in a 4.5-mile wind and make 87 revolutions per minute in a 25-mile wind with no load. Fig. 49 shows the horsepower of this mill for the four loads; also the maximum horsepower.

Mill No. 52.—This is a 14-foot Challenge wooden power mill on a 45-foot wooden tower, manufactured by the Challenge Windmill Company, of Batavia, Illinois. (See fig. 50.) It is a sectional mill, and has two side wheels for keeping the main wheel in the wind. The wind wheel has 102 sails, each 51.5 by 5 by 1.75 inches, set at an angle of 39° to the plane of the wheel.   The mill works a sheller, a grinder, and a pump.   There are two horizontal shafts, one of which works the grinder and sheller, the other the pump. The shaft that works the pump is 12 feet long and 1.5 inches in diameter; it is geared forward 1.5 to 1. The shaft that works the grinder is 6 feet long and 1.5 inches in diameter; it is geared forward 15.25 to 1. The well is a drilled well, 192 feet deep. The lift was 180 feet, the discharge 0.25 quart per stroke. The water is pumped into a large box, and passes to watering troughs when needed. The pump has a counterweight which raises on the downstroke and assists in lifting the water on the upstroke. The mean tem-

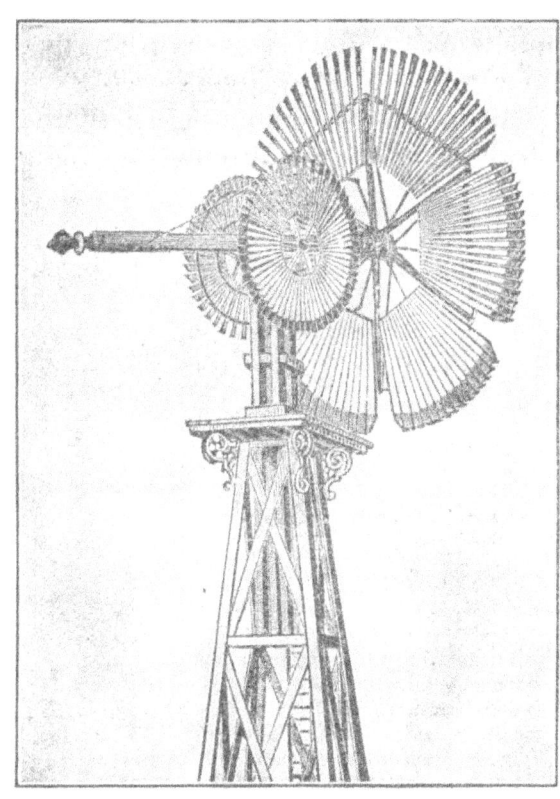

FIG. 50.—Mill No. 52—14-foot wooden Challenge.

perature was 92° F., the mean barometric pressure 28.6 inches.   The results of the tests are as follows:

Results of tests of mill No. 52—14-foot wooden Challenge.

Load on brake.	Load per revolution of wind wheel.	Number of revolutions of wind wheel per minute at given wind velocities (per hour).					Horsepower of mill at given wind velocities (per hour).				
		8 miles.	12 miles.	16 miles.	20 miles.	25 miles.	8 miles.	12 miles.	16 miles.	20 miles.	25 miles.
Lbs.	*Ft.-lbs.*										
0......	0	7	18	25	30	....	....	....	....	....	....
Pump	....		14	22	27	....	....	0.059	0.093	0.115	....
1......	432	....	....	17	22	....	....	....	0.222	0.287	....
2......	864	....	....	8	16	....	....	....	0.210	0.420	....

The revolutions of wind wheel per minute for loads of 0, 1, and 2 pounds are shown in fig. 51. For 0 load the pump shaft was running with the pump detached; the grinder shaft was not working. When pumping the grinder was not working. When the brake loads of 1 and 2 pounds were being used the pump shaft was not working. A curve was obtained giving the speed of the wind wheel for no brake load with the grinder shaft working and with the pump shaft not working. This curve nearly coincided with that for the pump load, showing that the friction of the grinder shaft was about equal to the pump load. With the pump shaft working, but not the pump, the mill will be seen to require a 7-mile wind to start it, and it makes only 30 revolutions in a 20-mile wind. Fig. 52 shows the horsepower of this mill.

This is a hard-running mill; there is too much friction. The side wheels do not respond to changes in the direction of the wind as quickly

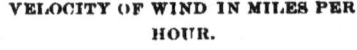

VELOCITY OF WIND IN MILES PER HOUR.

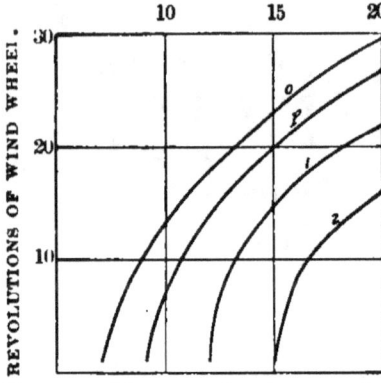

FIG. 51.—Diagram showing revolutions of wind wheel of mill No. 52—14-foot wooden Challenge. Curve marked 0 is for no brake load; curve P is for pump load; curves 1 and 2 are for brake loads of 1 and 2 pounds, respectively.

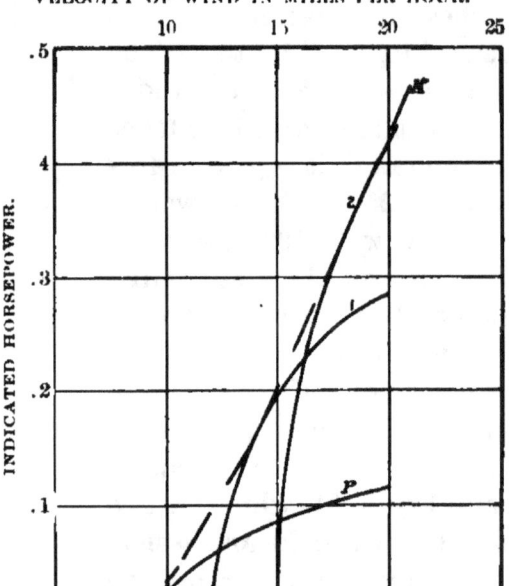

VELOCITY OF WIND IN MILES PER HOUR.

FIG. 52.—Diagram showing horsepower of mill No. 52—14-foot wooden Challenge. Curves marked 1 and 2 are for brake loads of 1 and 2 pounds, respectively; curve P is for pump load; dotted curve DK shows maximum power.

as does the vane or rudder in other mills. The wind exposure was very good and the mill was nearly new.

Mill No. 53.—This is a 12-foot Ideal power mill on a 33-foot wooden tower. The wind wheel has 21 curved sails, each 43.25 by 16.5 by 8.25 inches, set at an angle of 32° to the plane of the wheel. The horizontal shaft is geared forward 6.07 to 1. The mill had been in use about four years for shelling and grinding corn. Three brake loads were used in the test—0, 1.5, and 2.5 pounds. The mean temperature during the tests was 92° F., the mean barometric pressure 28.7 inches. The results of the tests are as follows:

Results of tests of mill No. 53—12-foot Ideal.

Load on brake.	Load per revolution of wind wheel.	Number of revolutions of wind wheel per minute at given wind velocities (per hour).					Horsepower of mill at given wind velocities (per hour).				
		8 miles.	12 miles.	16 miles.	20 miles.	25 miles.	8 miles.	12 miles.	16 miles.	20 miles.	25 miles.
Pounds.	*Ft.-lbs.*										
0	0	25	40	51	60	68	......	0.239	0.338	0.420	0.500
1.5	272	......	29	41	51	60	......	......	0.440	0.606	0.745
2.5	455	......	......	32	44	54	......	......			

Fig. 53 shows the number of revolutions per minute of the wind wheel for the three brake loads. The mill will be seen to start in about a 5-mile wind and to make 69 revolutions a minute in a 25-mile wind with no load. Fig. 54 shows the horsepower for the several loads.

Mill No. 54.—This is a 12-foot Aermotor like No. 27, on a 47-foot tower. The 12-foot Aermotor No. 27 was found to be so much greater in power and speed than other 12-foot mills tested that we thought

VELOCITY OF WIND IN MILES PER HOUR.

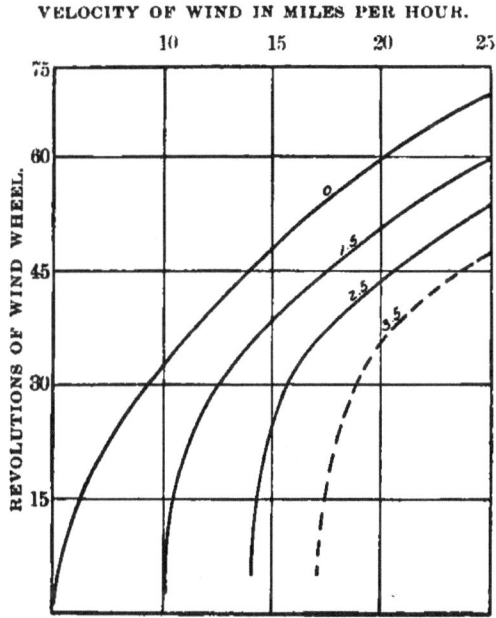

FIG. 53.—Diagram showing revolutions of wind wheel of mill No. 53—12-foot Ideal. Curves marked 0, 1.5, 2.5, and 3.5 are for brake loads of 0, 1.5, 2.5, and 3.5 pounds, respectively.

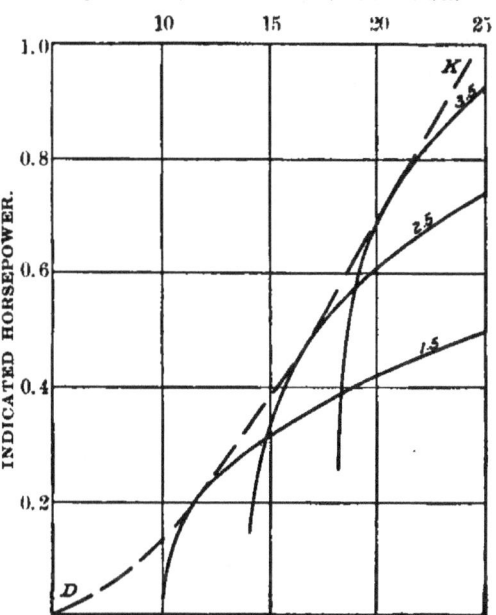

FIG. 54.—Diagram showing horsepower of mill No. 53—12-foot Ideal. Curves marked 3.5, 2.5, and 1.5 are for brake loads of 3.5, 2.5, and 1.5 pounds, respectively; dotted curve *DK* shows maximum power.

it wise to test another of the same make and size under somewhat different conditions. No. 54 had been in use about two years. Two brake loads were used—0 and 297 foot-pounds per revolution of wind wheel. The mean temperature during the test was 92° F., the mean barometric pressure 29.2 inches. The results of the tests are as follows:

Results of tests of mill No. 54—12-foot Aermotor.

Load on brake.	Load per revolution of wind wheel.	Number of revolutions of wind wheel per minute at given wind velocities (per hour).					Horsepower at given wind velocities (per hour).				
		8 miles.	12 miles.	16 miles.	20 miles.	25 miles.	8 miles.	12 miles.	16 miles.	20 miles.	25 miles.
Pounds. 0 1.75	*Ft.-lbs.* 0 297	31	49 38	63 52	75 63	87 75	....	...... 0.34	...... 0.46	...... 0.57	...... 0.68

It will be seen that the results of the tests of this mill apparently agree closely with those of No. 27, page 86. The agreement, however, is not so close as it appears, since the temperature for No. 54 is much higher. It is, however, about the same as the temperature for other mills, so that we can still use the results found for mill No. 27 in comparing its power and speed with those of other mills.

COMPARISON OF POWER MILLS.

Comparison of 12-foot Ideal (No. 53) with 14-foot Ideal (No. 34).—It must be remembered that in this and in all other comparisons no correction is made for difference in temperature and barometric pressure. The speeds for no load and the maximum horsepowers for these mills are as follows:

Comparison of results for 12-foot Ideal and 14-foot Ideal.

Mill.	Number of revolutions of wind wheel per minute at given wind velocities (per hour).					Maximum horsepower at given wind velocities (per hour).				
	8 miles.	12 miles.	16 miles.	20 miles.	25 miles.	8 miles.	12 miles.	16 miles.	20 miles.	25 miles.
12-foot Ideal	25	40	51	60	68	0.06	0.23	0.44	0.70	......
14-foot Ideal	20	34	44	53	58	0.06	0.22	0.46	0.74	......
Ratio of circumference velocities (12/14)	1.07	1.00	0.99	0.97	1.00	......	......	......	......	......

It will be seen that the useful power of the 14-foot mill is very little more than that of the 12-foot mill, and that the circumference velocities are nearly the same for no load, where no horizontal shaft is being turned. In the case of the 12-foot mill the brake was on the foot gear and there was no horizontal shaft to turn, but in the case of the 14-foot mill the brake pulley was on a shaft 15 or 20 feet long. We believe that if the brake pulley had been on the foot gear and the line shaft thrown out of gear, so as to eliminate shaft friction, the mill would have shown at least 10 per cent more power. The tower obstructs the wheel somewhat and reduces the power.

Comparison of 12-foot Aermotor (No. 27) with 14-foot Ideal (No. 34).—The speeds for no load and the maximum horsepowers for these mills are as follows:

Comparison of results for 12-foot Aermotor and 14-foot Ideal.

Mill.	Number of revolutions of wind wheel per minute at given wind velocities (per hour).					Maximum horsepower at given wind velocities (per hour).			
	8 miles.	12 miles.	16 miles.	20 miles.	25 miles.	8 miles.	12 miles.	16 miles.	20 miles.
12-foot Aermotor	30	49	63	75	87	0.09	0.33	0.66	1.05
14-foot Ideal	20	34	44	53	58	0.06	0.22	0.46	0.74
Ratio of circumference velocities (12/14)	1.29	1.23	1.22	1.21	1.28	......	......	......	......

It will be seen that the number of revolutions per minute of the 12-foot mill is 50 per cent greater than for the 14-foot mill. This is true for the lower as well as for the higher velocities, where the governing of the mill does not enter to reduce the speed. It will be seen also that the 12-foot mill is producing from 42 to 50 per cent more horsepower than the 14-foot mill. The temperature was 49° higher and the pressure 0.6 inch lower when the 14-foot mill was tested than when the 12-foot mill was tested. The effect is to lessen the difference between the power and speeds of the mills.

Comparison of 12-foot Aermotor (No. 27) with 16-foot Aermotor (No. 44).—Attention has already been drawn to the similarity between the speed curves of these mills (figs. 35 and 42) and between the power curves. (The power curves for mill No. 44 are shown in fig. 43; those for mill No. 27 were platted, but are not published because of lack of space.) If we compare the number of revolutions per minute for no load, we shall see that they are to each other nearly inversely as the diameter, or that the circumference velocities of the two mills are the same in all wind velocities. We may compare the brake horsepower and the speed as follows:

Comparison of results for 12-foot Aermotor and 16-foot Aermotor.

Mill.	Number of revolutions of wind wheel per minute at given wind velocities (per hour).					Maximum horsepower at given wind velocities (per hour).		
	8 miles.	12 miles.	16 miles.	20 miles.	25 miles.	10 miles.	15 miles.	20 miles.
12-foot Aermotor	30	49	63	75	87.0	0.21	0.58	1.05
16-foot Aermotor	23	38	48	56	64.5	0.29	0.82	1.55
Ratio of circumference velocities ($\frac{16}{12}$)	1.02	1.03	1.02	0.99	0.99	1.38	1.41	1.48

From this it will be seen that the power of the 16-foot Aermotor is about 1.22 times that of the 12-foot. The ratio of the squares of the diameters is 1.78; the ratio of the diameters 1.33. It will be seen that the power does not increase as the squares of the diameters, as is often stated; it increases faster than as the diameters, but more nearly as the diameters than as the squares of the diameters.

Comparison of 16-foot Althouse wooden mill (No. 28) with 16-foot Aermotor (No. 44).—From the following table it will be seen that the wind wheel of No. 44 is revolving from 56 to 77 per cent faster than the wind wheel of No. 28. The latter, however, has to overcome the friction of 10 or 12 feet of line shafting. It will be seen that No. 44 is yielding from 70 to 167 per cent more power than No. 28. The superiority of the steel mill over the wooden mill is very evident in this case.

Comparison of results for 16-foot Althouse and 16-foot Aermotor.

Mill.	Number of revolutions of wind wheel per minute at given wind velocities (per hour).					Maximum horsepower at given wind velocities (per hour).			
	8 miles.	12 miles.	16 miles.	20 miles.	25 miles.	8 miles.	12 miles.	16 miles.	20 miles.
16-foot Althouse................	13	23	30	36	40.0	0.06	0.29	0.52	......
16-foot Aermotor...............	23	38	48	56	64.5	0.16	0.48	0.93	1.55
Ratio of circumference velocities $\left(\frac{steel}{wood}\right)$	1.77	1.65	1.60	1.55	1.61	2.67	1.65	1.78	......

Comparison of 22.5-foot Halliday wooden mill (No. 49) with 16-foot Aermotor (No. 44).—From the following table it will be seen that the steel mill makes about two revolutions to one revolution of the wooden mill, and that its power is from 41 to 167 per cent greater.

Comparison of results for 22.5-foot Halliday and 16-foot Aermotor.

Mill.	Number of revolutions of wind wheel per minute at given wind velocities (per hour).				Maximum horsepower at given wind velocities (per hour).			
	8 miles.	12 miles.	16 miles.	20 miles.	8 miles.	12 miles.	16 miles.	20 miles.
22.5-foot Halliday............	11	20	25	30	0.06	0.34	0.62	0.88
16-foot Aermotor.............	23	38	48	56	0.16	0.48	0.93	1.55
Ratio of circumference velocities $\left(\frac{16}{22.5}\right)$	1.48	1.35	1.36	1.32	2.67	1.41	1.50	1.76

Comparison of 12-foot Aermotor (No. 27) with 22.5-foot Halliday wooden mill (No. 49).—Comparing the power of these mills, we have the following:

Comparison of results for 12-foot Aermotor and 22.5-foot Halliday.

Mill.	Number of revolutions of wind wheel per minute at given wind velocities (per hour).				Maximum horsepower at given wind velocities (per hour).			
	8 miles.	12 miles.	16 miles.	20 miles.	8 miles.	12 miles.	16 miles.	20 miles.
12-foot Aermotor.............	30	49	63	75	0.09	0.33	0.66	1.05
22.5-foot Halliday.............	11	20	25	30	0.06	0.32	0.63	0.94
Ratio of circumference velocities $\left(\frac{12}{22.5}\right)$	1.46	1.31	1.34	1.32	1.50	1.03	1.05	1.12

It will be seen that this 22.5-foot wooden mill does not furnish as much power as a good 12-foot steel mill.

Comparison of wooden power mills.—Of the mills in the following table the 12-foot has the least friction; the tower being in front of the windmill obstructs the wheel and reduces the power. The 14-foot mill probably has more friction than the others. The 16-foot Irrigator has too few sails; with more sails the power could probably be increased 75 per cent or more.

Comparison of results of tests of wooden power mills.

Mill	Number of revolutions of wind wheel per minute at given wind velocities (per hour).					Horsepower at given wind velocities (per hour).			
	8 miles.	12 miles.	16 miles.	20 miles.	25 miles.	8 miles.	12 miles.	16 miles.	20 miles.
12-foot Monitor (No. 50)......	16	33	44	54	64	0.02	0.10	0.23	0.38
14-foot Challenge (No. 52)....	7	18	25	30	......	0.01	0.10	0.25	0.42
16-foot Althouse (No. 28).....	13	23	30	36	40	0.06	0.29	0.52	0.84
16-foot Irrigator (No. 30).....	12	25	32	41	44	0.02	0.16	0.30	0.44
22.5-foot Halliday (No. 49)....	11	20	25	30	......	0.06	0.32	0.63	0.94

It will be seen that the 16-foot mill (No. 28) is furnishing from 2.2 to 2.9 times more useful power than the 12-foot mill (No. 50). It will also be seen that the power of this 16-foot mill compared with that of the 12-foot mill increases faster than as the squares of the diameters, while the power of the 22.5-foot mill compared with that of the 12-foot mill does not increase as fast as the squares of the diameters, and the power of the 22.5-foot mill compared with that of the 16-foot mill does not increase as fast as the first power of the diameters.

Comparison of 12-foot Monitor wooden mill (No. 50) with 12-foot Aermotor (No. 27) and with 12-foot Ideal (No. 53).— Fig. 55 shows the speed, in revolutions, of the

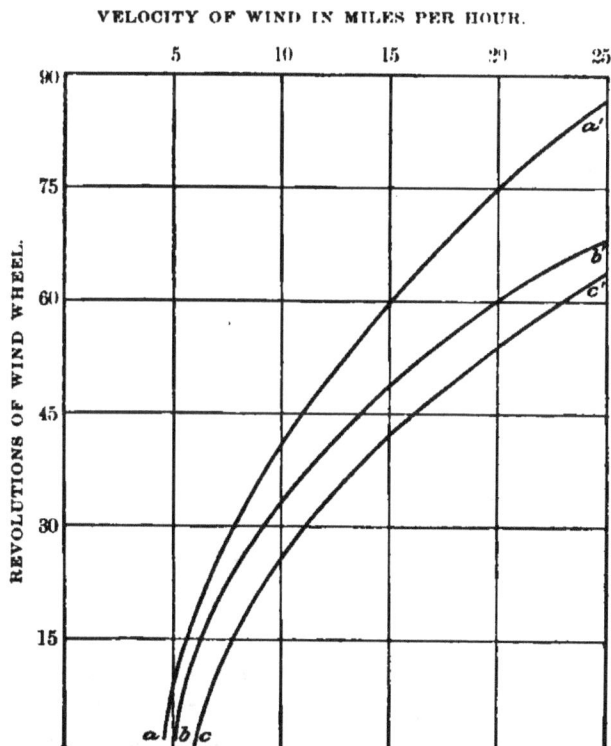

FIG. 55.—Comparative diagram of revolutions of wind wheels of mills Nos. 50, 53, and 27. Curve *aa'* is for 12-foot Aermotor; *bb'* is for 12-foot Ideal; *cc'* is for 12-foot Monitor.

wind wheels of these mills for no useful load. The friction is small in each case. The Monitor has a swivel gearing, and the Ideal has a ball gearing to carry the weight of the shaft, both of which reduce the friction. It is fair to say that the friction load of the Aermotor is at least as great as that of the other mills. It will be seen that the speed of the Ideal is noticeably greater than that of the Monitor; and that the speed of the Aermotor is much greater than that of the Ideal, especially for high wind velocities. This is the reason the Aermotor is so much more powerful than other mills. It revolves much faster, and the power is directly proportional to the speed. But why is its speed greater for the same load?

Comparison of results for 12-foot Monitor, 12-foot Ideal, and 12-foot Aermotor.

Mill.	Load per revolution of wind wheel.	Number of revolutions of wind wheel per minute at given wind velocities (per hour).					Horsepower at given wind velocities (per hour).				
		8 miles.	12 miles.	16 miles.	20 miles.	25 miles.	8 miles.	12 miles.	16 miles.	20 miles.	25 miles.
	Ft.-lbs.										
12-foot Monitor	0	16	33	44	54	64					
	120		25	35	43	52		0.091	0.127	0.156	0.189
	314			27	34	40			0.230	0.324	0.381
12-foot Ideal	0	25	40	51	60	68					
	272		39	41	51	60		0.239	0.338	0.420	0.500
	455			32	44	54			0.440	0.606	0.745
12-foot Aermotor	0	30	45	63	75	87					
	222	16	43	57	70	81	0.089	0.285	0.386	0.458	0.523
	444		23	48	65	77		0.303	0.653	0.890	1.020
	666			12	50	72			0.234	1.028	1.451

The dimensions of the principal parts of the wind wheels of these mills and the mean temperature and pressure when the tests were made are as follows:

Dimensions of principal parts of mills Nos. 50, 53, and 27.

Mill.	Number of sails.	Dimensions of sails.	Angle of sails.	Gearing.	Mean temperature.	Mean barometric pressure.
		Inches.	°		° *F.*	*Inches.*
12-foot Monitor (No. 50)	96	44 x 4¼ x 1¼	34	3.66 : 1	83	28.6
12-foot Ideal (No. 53)	21	43½ x 16¼ x 8¼	32	6.07 : 1	92	28.7
12-foot Aermotor (No. 27)	18	44 x 18½ x 7½	31	6.00 : 1	46	28.9

The temperature when the 12-foot Aermotor was tested was much lower than when the other two mills were tested, but, as already stated, the temperature when Aermotor No. 54 was tested was 92° F., and it showed a speed and power about the same as Aermotor No. 27; so we may leave this difference in temperature out of account.

It will be seen that the Monitor is not geared forward as much as the other two, but this does not affect the speed for no load. The sail angle is about the same for all, also the length of sail; but the number of sails and the width are quite different. The Aermotor has few sails, but of large size; the Ideal has more and somewhat smaller sails; the Monitor has many small sails, and its tower is located in front of the wind wheel. Its sails are plane, instead of curved, all of which tends to decrease its power.

Fig. 56 shows the curves of maximum horsepower. It will be seen that the difference between the horsepower of the Aermotor and that of the Ideal is about the same as the difference in their speeds for the same wind velocity, but the difference between the power of the Monitor and that of the Ideal is much greater than the corresponding difference in their speeds. The wooden mill, therefore, not only has a less speed at a given wind velocity than the steel mill, but it carries a proportionately less load. For example, in a 20-mile wind a load of 120 foot-pounds per revolution reduces the speed of the Monitor from 54 to 43 (or 11) revolutions per minute, while in the case of the Aermotor a load of 220 foot-pounds (about 80 per cent greater) reduces the speed from 75 to 70

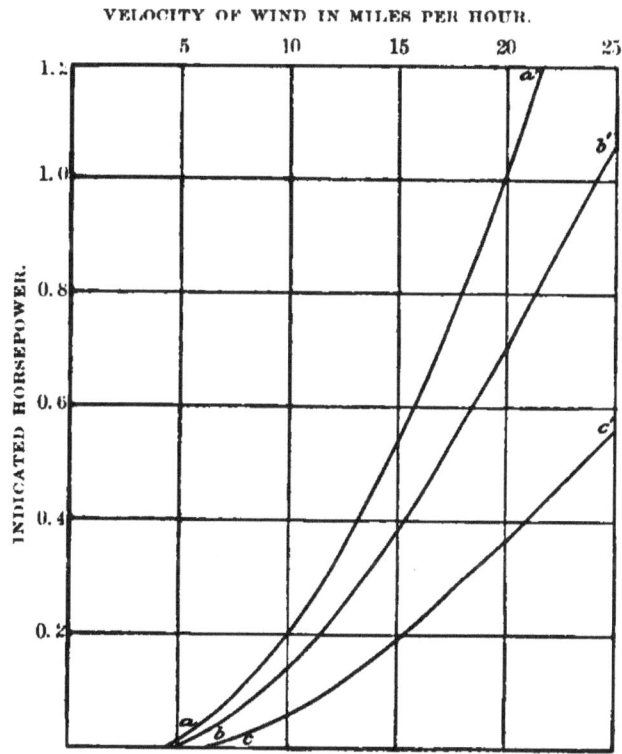

FIG. 56.—Comparative diagram of horsepower of mills Nos. 50, 53, and 27. Curve aa' is for 12-foot Aermotor; bb' is for 12-foot Ideal; cc' is for 12-foot wooden Monitor.

(only 5) revolutions per minute. In the case of the Ideal a load of 272 foot-pounds per revolution (2.3 times the load of the Monitor) reduces the speed from 60 to 51 (or 9) revolutions per minute.

COMPARISON OF PUMPING MILLS WITH POWER MILLS.

Comparison of 12-foot pumping mill (No. 3) with 12-foot power mill (No. 27).—The load per stroke of No. 3 (see page 30, Part I) is 415.3 foot-pounds. The wind wheel makes 3.3 revolutions to 1 stroke of the pump, so that the load per revolution of wind wheel is 124.5 foot-pounds. This is less than the smallest load used in testing No. 27, viz, 222 foot-pounds per revolution. A diagram was platted showing the useful work, in horsepower, of these mills for these loads. The curve for the pumping mill was seen to start with a little less wind velocity than that of the power mill, indicating a somewhat less total load.

Comparing the ordinates of these curves for different wind velocities, the following ratios were obtained, which give, approximately, the pump efficiency, no allowance being made for difference in temperature and pressure:

8	12	16	20	25	30
0.60	0.53	0.54	0.54	0.58	0.58

The mean of these ratios is 0.56. If the useful load of the pumping mill had been somewhat greater, so that the mills would have started at the same wind velocity, the ratio, or pump efficiency, would be about 60 per cent, which is about what might be expected of this pump under this lift. The ratio of the useful loads is $125 \div 222 = 0.57$. This ratio would probably be about 0.60 if the loads were such that the mills would start at the same wind velocity.

Comparison of 16-foot pumping Aermotor (No. 9) with 16-foot power Aermotor (No. 44).—The useful load of No. 9 (see pages 36 to 37, Part I) is 1,013 foot-pounds per stroke of pump, or 304 foot-pounds per revolution of wind wheel. The smallest useful load of No. 44 is 528 foot-pounds per revolution of wind wheel. A diagram was platted showing the useful horsepower of these mills for these loads. For this particular load (1,013 foot-pounds) the pumping mill was seen to start at a somewhat less wind velocity than the power mill, indicating that the total load of the pumping mill was somewhat less than that of the power mill. The ratio of any two of the ordinates gave, approximately, the pump efficiency for that wind velocity, the difference in temperature and pressure being neglected.

These ratios for four velocities are as follows:

12	16	20	25
0.72	0.67	0.69	0.63

If the pump load had been somewhat greater—such that the mills would start at the same wind velocity—the mean ratio would be about 70 per cent. This, again, is about what is expected for the efficiency of this pump, which is somewhat better than mill No. 3 and has a higher lift. The ratio of the useful loads of these mills is $304 \div 528 = 0.57$. This ratio would probably be about 60 per cent if the mills started at the same wind velocity.

It is interesting to compare the performance of these mills still further. The speeds of the wheels and the horsepowers are as follows:

Comparison of results for 16-foot pumping Aermotor and 16-foot power Aermotor.

Mill.	Number of revolutions of wind wheel per minute at given wind velocities (per hour).				Horsepower at given wind velocities (per hour).				Load per revolution of wind wheel.
	12 miles.	16 miles.	20 miles.	25 miles.	12 miles.	16 miles.	20 miles.	25 miles.	
16-foot pumping mill (No. 9)	31	42	52	59	0.325	0.433	0.548	0.601	Ft.-lbs. 0.304
16-foot power mill (No. 44)	28	41	50	58	0.45	0.65	0.80	0.95	0.528
Ratio of pump power to mill power					0.72	0.67	0.69	0.63	0.58

It will be seen that the wind wheel of the pumping mill is making from one to two more revolutions per minute than that of the power mill.

Comparison of 22.5-foot pumping mill (No. 36) with 22.5-foot power mill (No. 49).—In this comparison we will use the curve of 5 pounds, or 1,087 foot-pounds, per revolution of wind wheel as the speed for this load, corresponding more nearly with that for the pump load than any other. The speeds of the mills and the horsepowers are as follows:

Comparison of results for 22.5-foot pumping mill and 22.5-foot power mill.

Mill.	Number of revolutions of wind wheel per minute at given wind velocities (per hour).				Horsepower at given wind velocities (per hour).				Load per revolution of wind wheel.
	12 miles.	16 miles.	20 miles.	25 miles.	12 miles.	16 miles.	20 miles.	25 miles.	
22.5-foot pumping mill (No. 36)	12	17	20	24	0.090	0.124	0.150	0.182	Ft.-lbs. 0.248
22.5-foot power mill (No. 49)	11	18	22	26	0.342	0.598	0.724	0.856	1.087
Ratio of pump power to mill power					0.26	0.20	0.20	0.22	0.23

The efficiency of the pump is, therefore, not more than 22 per cent. The ratio of useful loads is 22 per cent. (For description of pumping mill No. 36, see pages 52 to 53, Part I.)

EFFECT OF TENSION OF SPRING ON SPEED AND HORSEPOWER OF MILL.

The effect of tightening the spring which holds the wind wheel of mill No. 18 in the wind has been shown on page 44, Part I. The effect of a reduction in the tension of the spring of the 16-foot power mill

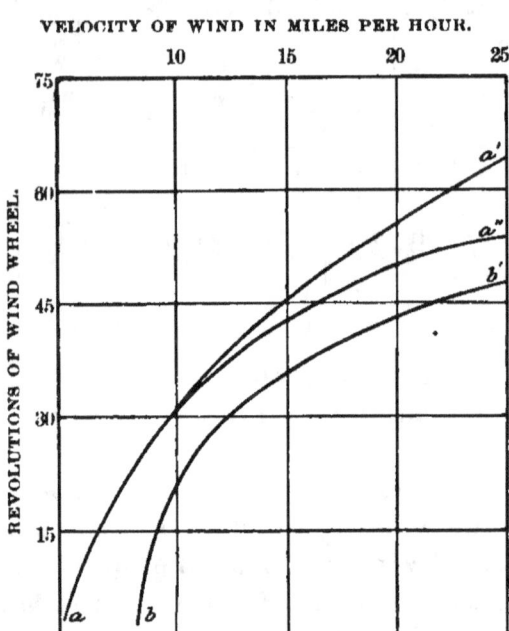

VELOCITY OF WIND IN MILES PER HOUR.

FIG. 57.—Diagram showing effect of tension of spring of mill No. 44—16-foot Aermotor. Curve *aa'* is for no load, spring new and stiff; *aa''* is for no load, spring relaxed, mill having been out of use eight months; *bb'* is for 3-pound brake load with relaxed spring.

(No. 44) is shown in fig. 57. The curve *aa'* is for no load and the spring new and stiff. The curve *aa''* is for the same load (none) and spring after the mill had been out of use about eight months. It will be seen that these curves coincide up to a velocity of about 10 miles an hour, after which they separate rapidly. In a 25-mile wind the number of revolutions of the wind wheel per minute has been reduced from 64.5 to 54 by the decrease in the tension of the spring. The curve *bb'* is for a 3-pound brake load with relaxed spring. It will be seen to be nearly parallel to the curve *aa''*, showing that the effect of the load when the spring is relaxed is similar to that when it is taut. We see how important a factor tension of spring is on the power of a mill. It also shows that if a spring is to be used in place of a weight there should be some easy way to change its tension.

MATHEMATICAL DISCUSSION OF TESTS OF TWO AERMOTORS.

In this discussion the wind velocities are those found by the use of the Robinson cup anemometer. A comparison of these velocities with true wind velocities will be given later. The Aermotors are selected for this discussion because their power is greater than that of any other form of mill that we have tested, and because their speed and power curves are derived from a greater number of observations than those of any other mill.

Discussion of tests of 12-foot Aermotor (No. 27).—The curves showing the number of revolutions of wind wheel (called speed curves) of this mill for four brake loads are given on page 88—fig. 35. Each of these curves is seen to resemble a parabola the axis of which is the x coordinate axis on which the wind velocities are marked. Each of these has the form $y^2 = a + bx$, in which x is the wind velocity in miles per hour, y the speed of wind wheel in revolutions per minute, and a and b constants.   For the curve of no load (0) we have $y = 50$ when $x = 12$, and $y = 75$ when $x = 20$.   Hence we have

$$50^2 = a + 12b, \text{ and}$$
$$75^2 = a + 20b.$$

Solving these, for a and b we have $a = -2,187$ and $b = 391$; and the equation of the curve is

$$y^2 = -2,187 + 391x. \tag{1}$$

For the speed curve of 2 pounds we see that $y = 43$ when $x = 12$, and that $y = 70$ when $x = 20$.   Hence we have

$$43^2 = a + 12b, \text{ and}$$
$$70^2 = a + 20b.$$

Solving these, we have $a = -2,728$ and $b = 381$; and we have the equation of this 2-pound curve

$$y^2 = -2,728 + 381x. \tag{2}$$

Proceeding in a similar way, we have for the equation of the 4-pound curve

$$y^2 = -4,384 + 424x. \tag{3}$$

For the equation of the 6-pound curve we have

$$y^2 = -9,400 + 595x. \tag{4}$$

The speed as determined by measurement and as found from these equations for several wind velocities is shown the following table.

The starting velocities are found by making $y = 0$ and solving for x in equations 1 to 4.

Table showing revolutions per minute of 12-foot Aermotor (No. 27) under different loads and at different wind velocities.

Wind velocity per hour.	No load.		2-pound load.		4-pound load.		6-pound load.	
	Meas-ured.	Com-puted.	Meas-ured.	Com-puted.	Meas-ured.	Com-puted.	Meas-ured.	Com-puted.
	Rev.	Rev.	Rev.	Rev.	Rev.	Rev.	Rev.	Rev.
8 miles	30	30	16	18				
12 miles	49	50	43	43	23	27		
16 miles	63	64	57	58	48	49	12	10
20 miles	75	75	70	70	65	64	50	50
25 miles	87	87	81	82	77	78	72	74
Starting velocity	4.5	5.6	7.0	7.1	10.2	10.3	15.3	15.8

If the origin of coordinates for equation 1 be moved to the point where the curve crosses the axis of x, the equation will then be of the form $y^2 = 391x'$, from which we have $y = \sqrt{391x'}$; that is, the speed for a constant load increases as the square root of the wind velocity.

The close agreement between the measured and computed speeds, especially for the curve of no load, is noticeable. The measured and computed starting velocities differ somewhat. This was expected, since it is difficult to get the starting velocities from observation.

Hereafter in this discussion the computed instead of the observed speeds and starting velocities will be used. It must be remembered, however, that these are not what may be called theoretical results. They are obtained from measurements, not from theory, and are the adjusted values of the observed quantities.

The power curves for this mill for three brake loads were platted, but are not published because of lack of space. The curves are parabolas with their axes horizontal. This follows at once from the fact that the corresponding speed curves are parabolas. The power is proportional to the product of the load and speed. When the load is constant, as it is for one of these speed curves, the power varies as the speed, and hence the load curves are parabolas.

The equation of any one of the curves—as, for example, the 2-pound curve—may be found as follows: The formula for horsepower is H. P. $= 2\pi RuL \div 33,000$, $R = 35.5 \div 12$ and $L = 2$. Hence H. P. $= 2 \times \overline{22 \div 7} \times 6 \times 35.5 \times 2 \times u \div \overline{12 \times 33,000} = 0.0067u = Ku$ where $K = 0.0067$.

In the speed equations u is what we have called y, and $y = -2,728 + 381x$. Hence—

H. P. $= Ky = K\sqrt{-2,728 + 381x}$, and
(H. P.)$^2 = K^2(-2,728 + 381x) = -0.1225 + 0.017x$.   (5)

In the diagram of power curves, platted but not reproduced here, the curve of maximum power was found to resemble a parabola the axis of which was vertical with its vertex on the y coordinate axis below the origin. The form of its equation is $x^2 = a + by$, x being the wind velocity, in miles per hour, y the horsepower, and a and b constants. For $x = 5$, $y = 0$, and for $x = 20$, $y = 1.05$. Substituting these values in the above equation we have: $25 = a$, and $400 = a + 1.05b$. From these we have: $a = 25$, $b = 357$, and the equation of the maximum power curve is—

$$x^2 = 25 + 357y. \qquad (6)$$

For $x = 5, 10, 15, 20$, and 25, y has the values $0, 0.21, 0.56, 1.05$, and 1.69, which agree closely with those taken from the curve.

For the mill to yield the greatest amount of power possible the load should increase as the wind velocity increases. In an 8.5-mile wind a 2-pound load gives the maximum power; in a 14-mile wind a 4-pound load gives the maximum power, and in a 21-mile wind a 6-pound load gives the maximum power.

We wish to determine how the load and speed of the mill vary with

MURPHY.] MATHEMATICAL DISCUSSION. 113

the wind velocity for the curve of maximum power. The load curves were found to be tangent to the curve of maximum power for loads and velocities about as follows: The 0 curve is tangent at $x = 5$, the 2-pound curve at $x = 8.5$, the 4-pound curve at $x = 21$. For a constant increment of 2 pounds in the load the increment of wind velocity changed from 3.5 miles to 7 miles. Hence the velocity increases faster than the loading. For each of these four points on the curve the load and horsepower are known. Hence we can find the number of revolutions from the equation—

$$\text{H. P.} = 2 \times \pi \times R \times 6 \times 4 \times L \div 33,000. \qquad (7)$$

The wind velocities, loads, powers, and speeds for these four points of tangency are as follows:

Data regarding points of tangency of power curves with curves of maximum power of Aermotor No. 27.

Wind velocity per hour.	Load.	Horsepower.	Revolutions per minute.
	Pounds.		
5 miles.................	0	0	0
8.5 miles...............	2	0.13	19
14 miles................	4	0.50	38
21 miles................	6	1.15	57

The speeds, in revolutions, as here given are platted in fig. 35, giving the curve PQ, which is the speed curve for maximum power. The proper load for maximum power can now be found for any wind velocity from equation 7, the speed being taken from this speed curve. The ratio of the speed at maximum load to the speed at 0 load, for the wind velocities 10, 15, and 20 miles an hour, is 0.63, 0.70, and 0.74, respectively, showing quite an increase. The following table gives additional information in regard to speed and load for the curve of maximum power:

Data regarding speed and load for curve of maximum power of Aermotor No. 27.

Wind velocity per hour.	Load per revolution of wind wheel.	Revolutions per minute.	L − L'.	S − S'.	LS.	LS − L'S'.	Δ^2.
	Pounds.						
5 miles....	0	0	....	....	....	....	
10 miles....	2.1	25	2.1	25	52.5	52.5	....
15 miles....	4.0	41	1.9	16	160.4	107.9	55.4
20 miles....	5.9	55	1.9	14	324.5	164.1	56.2

The fourth column gives the differences between the successive loads, or the increments of loading. These are seen to decrease somewhat, showing that the load does not increase quite as fast as the wind velocity. The fifth column gives the differences between the successive speeds, and shows that the speed does not increase quite as fast as the wind velocity. The sixth column gives the products of the loads and speeds, which is proportional to the horsepower. The seventh column gives the increments of horsepower, the eighth column the difference between the figures in the seventh column. These, being nearly constant, show that the curve of maximum power is of the second degree.

The following table contains additional interesting information in regard to the speed of this mill:

Data in regard to speed of mill No. 27—12-foot Aermotor.

Wind velocity per hour.	Revolutions per minute, no load.	Circumference velocity, in miles, no load.	Ratio of circumference velocity to wind velocity, no load.	Revolutions per minute at maximum load.	Ratio of speed at maximum load to speed at no load.
8 miles	30	12.9	1.61	17	0.57
12 miles	49	21.0	1.75	32	0.65
16 miles	63	27.0	1.70	44	0.70
20 miles	75	32.1	1.60	54	0.72
25 miles	87	37.3	1.50		

The results obtained from this 12-foot mill may be stated as follows, in terms of cup anemometer velocities:

(1) The speed of the wheel for a constant load varies as the square root of the wind velocity.

(2) The speed of the wheel for maximum load increases slightly faster than the first power of the wind velocity.

(3) The power of the mill for a constant load varies as the square root of the wind velocity.

(4) The maximum power of the mill varies as the square of the wind velocity.

(5) The load for maximum power does not increase quite as fast as the wind velocity.

(6) The ratio of speed for maximum load to the speed for no load increases somewhat with the wind velocity.

Discussion of tests of 16-foot Aermotor No. 44.—The speed curves for this mill are shown in fig. 42. They are seen to resemble the parabolas with horizontal axis. The equation of each has the form $y^2=a+bx$, y being the speed in revolutions per minute, x the wind velocity in miles per hour, and a and b being constants for any curve. We see that for $x=12$, $y=38$, and that for $x=20$, $y=56$. Hence we have

$$38^2=a+12b, \text{ and}$$
$$56^2=a+20b.$$

Solving these equations, we have $a=-1,094$, $b=211.5$, and the equation of the no-load speed curve is

$$y=-1,094+211.5x. \qquad (8)$$

Proceeding in a similar way, we have for the equation of the 3-pound load speed curve

$$y^2 = -1,790 + 214.5x. \qquad (9)$$

For the 5-pound load we have

$$y^2 = -2,304 + 212x. \qquad (10)$$

For the 8-pound load we have

$$y^2 = -2,715 + 197x. \qquad (11)$$

The speed and starting velocities as computed from these equations and as found by measurement are as follows:

Speed and starting velocities for 16-foot Aermotor No. 44.

Wind velocity per hour.	No load.		3-pound load.		5-pound load.		8-pound load.	
	Meas-ured.	Com-puted.	Meas-ured.	Com-puted.	Meas-ured.	Com-puted.	Meas-ured.	Com-puted.
8 miles	23.0	24.0						
12 miles	38.0	38.0	28.0	28.0	13	15.0		
16 miles	48.0	48.0	41.0	40.0	33	33.0	16.0	20.0
20 miles	56.0	56.0	50.0	50.0	44	44.0	36.0	35.0
25 miles	64.5	65.0	59.0	60.0	54	54.0	47.0	47.0
Starting velocity	4.5	5.1	8.0	8.3	11	10.9	14.5	13.8

The computed values are seen to agree closely with the measured values, so that these speed curves are parabolas of the form $y = \sqrt{a + bx}$. The power curves shown in fig. 43 are parabolas of this form for the reason given for the corresponding case of the 12-foot Aermotor. The curve of maximum power to which these power curves are tangent is a parabola with its axis vertical. Its equation has the form $x^2 = a + by$. We may obtain the data for finding the value of a and b by observing that when $x = 10$, $y = 0.30$; and that when $x = 20$, $y = 1.55$. We have

$$10^2 = a + 0.30b, \text{ and}$$
$$20^2 = a + 1.55b.$$

Solving these, we have $a = 28$, $b = 240$, and the equation of the curve is

$$x^2 = 28 + 240y. \qquad (12)$$

The values of y for four values of x are $x = 8$, $y = 15$; $x = 12$, $y = 0.48$; $x = 16$, $y = 0.95$; and $x = 20$, $y = 1.55$. It will be seen that these values agree closely with the measured horsepower for these velocities. By making $x = 0$ in equation 12 we have $y = -0.125$. The vertex of this maximum power curve is at a distance 0.125 below the axis of x. If the origin of coordinates be changed to this point, equation 12 will take the form $x^2 = K y'$, K being a constant and y' the horsepower referred to the new origin. Hence we see that the maximum horsepower varies as the square of the wind velocity.

To find the variation of the speed and load for this curve DK, we notice that the 3-pound curve is tangent at $x=10.5$, the 5-pound curve at $x=14$, the 8-pound curve at $x=19$, and the 11-pound curve at $x=24$. The horsepower is known at these points, so that the speed can be found from equation 7.

The wind velocity, load, horsepower, and speed for each of these points are as follows:

Data regarding points of tangency of power curves with curves of maximum power of Aermotor No. 27.

Wind velocity per hour.	Load.	Horsepower.	Revolutions per minute.
	Pounds.		
5 miles	0	0	0
10.5 miles	3	0.33	21
14 miles	5	0.70	26
19 miles	8	1.40	33
24 miles	11	2.17	37

The speeds here found are platted in fig. 42, giving the curve PQ, which gives the speed of the wheel for the maximum load. This curve is seen to be a nearly straight line for velocities above 9 or 10 miles an hour. Hence we may say that the speed increases as the first power of the wind velocity for maximum power.

The load for any wind velocity can now be found from the formula H. P. $= \dfrac{176\, Lu}{33,000}$, the speed being taken from the speed curve PQ. Or the loads can be measured from the load curve RS, fig. 42. This load curve (RS) for maximum power is seen to be a straight line, showing that for wind velocities above 9 or 10 miles an hour the load for maximum power varies nearly as the first power of the wind velocity.

The following table contains some interesting facts in regard to the working of this mill:

Data in regard to speed of mill No. 44—16-foot Aermotor.

Wind velocity per hour.	No load.			Maximum load.			Ratio of speed at maximum load to speed at no load.
	Revolutions per minute.	Circumference velocity in miles.	Ratio of circumference velocity to wind velocity.	Revolutions per minute.	Circumference velocity in miles.	Ratio of circumference velocity to wind velocity.	
8 miles	23	13.2	1.67	15	8.6	1.08	0.65
12 miles	38	21.7	1.81	23	13.2	1.10	0.61
16 miles	48	27.4	1.71	29	16.6	1.04	0.60
20 miles	56	32.0	1.60	34	19.4	0.97	0.60
25 miles	64	36.6	1.46	38	21.6	0.86	0.60

It will be seen that the ratio of the circumference velocity of the wheel to the wind velocity increases to 12 miles an hour and then decreases.   In a 12-mile wind the circumference of the wheel is moving 1.81 times faster than the wind that drives it.   It will be seen also that the circumference velocity of the wheel when carrying the maximum load is about equal to that of the wind that drives it, and that the speed of the wheel when carrying a maximum load is about 39 per cent less than its speed when carrying no useful load.

ACTION OF AIR ON THE SAIL OF AN AERMOTOR.

It is not our purpose to discuss this action from a theoretical point of view, but to explain it from the observed and computed results of the 16-foot Aermotor.   Fig. 58 shows the concave surface of one sail of a 16-foot Aermotor in a nearly horizontal position as it moves downward; w represents the velocity of the wind, and v the circumference velocity of the sail.   The curve JAE, fig. 59, shows the outer end of the sail, and to, in fig. 60, shows the inner end.   The cords of these arcs, or the plane of the sail, makes an angle (JOP) of 30° with the plane of the wheel.   The point E, fig. 59, represents a particle of air as it comes in contact with the sail when the wheel is carrying its best load; EF represents the velocity of the wind, EH the velocity of this point of the sail; then EG, the other side of the parallelogram constructed on EH and EF, is the velocity of this particle of air over the sail; EG is not tangent to the sail.   The point A represents a particle of air as it comes in contact with the moving sail, when the mill is carrying no load; AC represents the relative velocity of the particle of air.   It will be seen that the air does not enter the sail tangent to it, but more nearly tangent for best load than for no load.

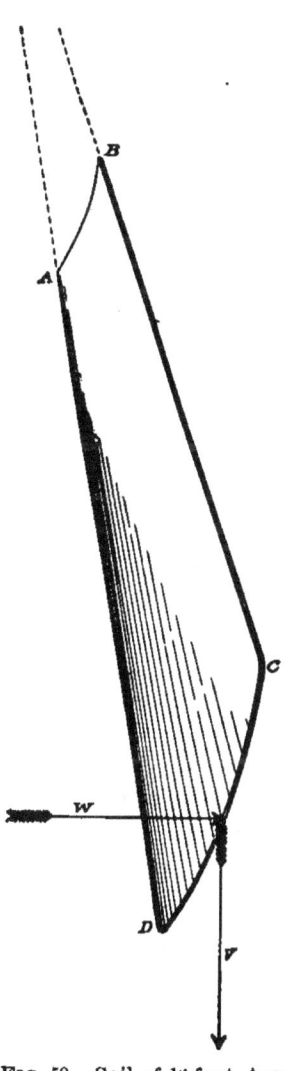

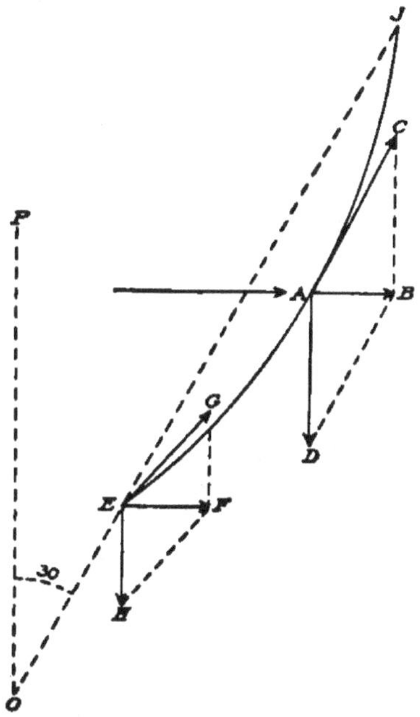

FIG. 58.—Sail of 16-foot Aermotor

FIG. 59.—Outer end of sail of 16-foot Aermotor.

In fig. 60 t represents a particle of air as it strikes the inner end of the sail when the mill is carrying no load, and a represents a particle of air as it strikes the inner end of the sail when the mill is carrying its maximum load. It will be seen that the air does not strike the sail tangent to it at any place for any load, but that it is most nearly tangent at the outer end of the sail at maximum load. In order for the air to enter the sail tangent to it at maximum load, the angle POJ should

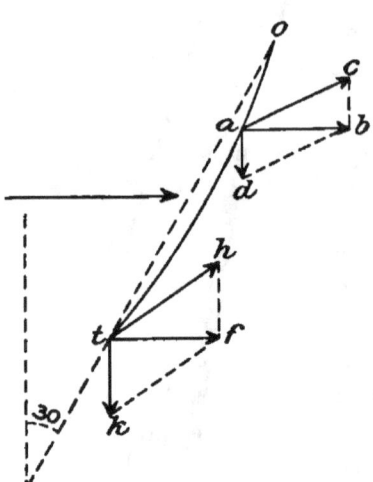

FIG. 60.—Inner end of sail of 16-foot Aermotor.

be a little greater than 30° at the outer end, and considerably greater than 30° at the inner end. As the load is decreased the angle POJ should be decreased.

USEFUL WORK OF TWO POWER MILLS IN A GIVEN TIME.

We can find the useful work of the 12-foot and the 16-foot Aermotors in a year, as we have for two pumping mills (pp. 69–71). For this purpose we will use the mean wind movement at Dodge, Kansas, from 1889 to 1893, as given by Mr. Willis L. Moore, Chief of Weather Bureau.[1] The mean number of hours per month that the wind velocity was 0 to 5, 6 to 10, etc. miles an hour at this place is given in the following table, also the mean horsepower of these two mills for each month. The number of horsepower hours for each mill each month is given at the bottom of the table. The horsepower hours for any month are found by multiplying the number of hours by the horsepower and adding the products.

Table showing useful work of 12-foot and 16-foot Aermotors in a year.

Month.	Mean wind movement at Dodge, Kansas, 1889–1893.							Total hours.	Horsepower hours.	
	0 to 5 miles.	6 to 10 miles.	11 to 15 miles.	16 to 20 miles.	21 to 25 miles.	26 to 30 miles.	31+ miles		12-foot mill.	16-foot mill.
	Hrs.	Hrs.	Hrs.	Hrs.	Hrs.	Hrs.	Hrs.			
January	200.9	253.0	156.2	74.4	37.2	14.9	7.4	744	250.5	358.0
February	176.0	230.1	128.6	74.4	40.6	20.3	6.8	677	251.6	361.5
March	126.5	208.3	178.6	119.0	59.5	29.8	22.3	744	386.6	558.9
April	115.2	172.8	158.4	115.2	72.0	43.2	43.2	720	461.2	671.3
May	119.0	193.5	171.1	119.0	74.4	37.2	29.8	744	433.9	629.4
June	122.4	187.2	136.8	108.0	86.4	50.4	28.8	720	452.0	658.2
July	141.4	215.8	178.6	119.0	59.5	22.3	7.4	744	339.8	489.3
August	178.6	230.6	156.3	96.7	59.5	14.9	7.4	744	297.5	427.6
September	165.6	180.0	151.2	93.6	72.0	36.0	21.6	720	379.6	550.5
October	208.3	230.6	141.4	74.4	22.3	14.9		744	294.0	423.5
November	194.4	266.4	129.6	64.8	36.0	14.4	14.4	720	244.9	350.9
December	186.0	267.9	141.4	81.8	44.6	14.9	7.4	744	262.2	374.6
Mean									337.8	487.8
Horsepower 16-foot mill		0.13	0.56	1.25	2.00	3.15				
Horsepower 12-foot mill		0.10	0.41	0.85	1.36	2.12				

[1] Some Climatic Features of the Arid Region, by Willis L. Moore. Washington, 1896.

MURPHY.]   MATHEMATICAL DISCUSSION.   119

The work done by these mills is greatest at this place in April (461 horsepower hours for the 12-foot and 671 horsepower hours for the 16-foot) and least in November (245 horsepower hours for the 12-foot and 351 horsepower hours for the 16-foot). The mean monthly power is 338 for the 12-foot mill and 488 for the 16-foot mill. Stating these results in another way, we may say that the 12-foot mill at this place will furnish on an average 1.3 horsepower 10 hours a day for 26 days a month, and the 16-foot mill will furnish 1.9 horsepower 10 hours a day for 26 days a month. It must be remembered that the wind velocity on the Great Plains is considerably greater than in the eastern part of the United States, and that consequently the horsepower hours of these mills when used in New York State, for example, will be considerably less than those given in the foregoing table.

MATHEMATICAL DISCUSSION OF TESTS OF JUMBO MILL NO. 55.

On page 46 we have given the results of tests of a 15.5-foot Jumbo mill working two 6-inch pumps. In order more fully to determine the power of this mill and its variation with the number and size of the sails, we have had constructed mill No. 55, shown in Pl. XVI, *B*. It is made of wood, the parts being fastened together with bolts. The diameter is 7.75 feet; length of sails, 11½ feet. There are 8 sails, each made of two boards 11½ feet long and 8 inches wide. There is no governor or other method of regulating the speed of the mill at high wind velocities, as in other mills, but there is a large door or shield on each side. By opening these the mill may be stopped. The mill is not fastened to the ground, but may be moved around by hand so that the wind strikes the sails at right angles. The shaft is 4 by 4 inches and 14 feet long, carefully turned down in a lathe to a diameter of 3 inches near each end. The friction brake is of wood, has an arm about 3½ feet long, and is made so as to fit on the end of the shaft. Oil was freely used on the brake. It was not found practicable to use loads greater than about 6 pounds on a 35-inch arm, as the friction burned the shaft; but the results for the four loads used showed that up to the maximum load the speed of the wheel, or the number of revolutions per minute, varies directly as the load, so that we can easily compute the load and speed for maximum power in any wind velocity. The weight of the wheel with its 8 sails was about 450 pounds. The coefficient of axle friction for well-oiled yellow pine is probably about 0.10, so that the axle friction was about 45 pounds. This weight (45 pounds), acting with a 1½-inch arm, is equivalent to about 2 pounds applied on the brake with a 35-inch arm. The friction on starting is probably 50 to 100 per cent greater than the friction of motion. The 0 brake load then really corresponds to a brake load of 2 or more pounds.

Four sets of tests were made of this mill, numbered 1, 2, 3, and 4. In the first set the full sail area of 8 sails, each 11½ feet by 16 inches,

was used. The number of revolutions of the wheel for the four brake
loads of 0, 1.75, 2.5, and 4.5 pounds was determined for velocities from
7 to 22 miles an hour. In the second set of tests there were 8 sails,
each having an area of 11½ feet by 8 inches; that is, each sail was
only half as wide as those used in the first set of tests. In the third
set of tests the sail area consisted of 4 sails, each 11½ feet by 16
inches; that is, every other full sail was removed. The fourth set of
tests was made to determine the effect of concentrating the air on the
sails and reducing the resistance due to air striking the shield and
glancing upward by the use of an inclined surface of approach to

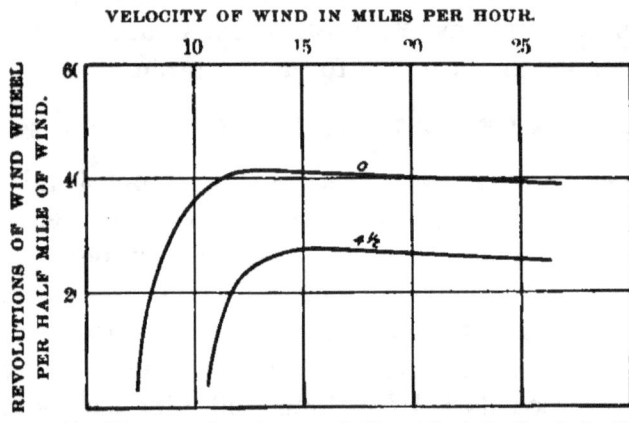

wheel. This inclined
surface had a length of
7 feet and formed an
angle of 11° with the
horizontal.

Test No. 1.—In Pl.
XVI, *B*, the mill is
shown with 8 sails, each
11½ feet by 16 inches.
Fig. 61 shows the num-
ber of revolutions of
the wheel per half mile
of wind for two loads—
0 and 4.5 pounds—on a
35-inch arm, for the

FIG. 61.—Diagram showing revolutions of wind wheel of mill
No. 55—7.75-foot Jumbo. Curve marked 0 is for no brake
load; curve marked 4½ is for a brake load of 4.5 pounds.

wind velocities shown. These curves for a mill without any means of
governing in high velocities are given for comparison with curves of
mills having a governor. It will be seen that these curves are nearly
horizontal straight lines beyond the point of maximum revolutions
per half mile. Thus, for no brake load the revolutions at 12 miles are
about 62, and at 25 miles about 59. In a mill with a governor, as, for
instance, that shown in fig. 36, the curve is more inclined, or the drop
in the number of revolutions is greater.

Fig. 62 shows the number of revolutions per minute for several
loads, fig. 63 shows the horsepower.

The results of these tests are as follows:

Results of tests of Jumbo mill No. 55 with 8 sails 11½ feet by 16 inches.

Load on brake.	Load per revolution of wind wheel.	Revolutions of wind wheel per minute at given wind velocities (per hour).					Horsepower at given wind velocities (per hour).				
		8 miles.	12 miles.	16 miles.	20 miles.	25 miles.	8 miles.	12 miles.	16 miles.	20 miles.	25 miles.
Lbs.	*Ft.-lbs.*										
0	0	11	24	33	41	49	......	......	......	......	......
1.75	32.0	......	22	30	38	45	......	0.022	0.030	0.036	0.043
2.5	45.8	......	20	28	36	43	......	0.028	0.039	0.050	0.060
4.5	82.5	......	17	25	33	40	......	0.043	0.063	0.082	0.100

From these results it is seen that the reduction in the number of revolutions per minute is proportional to the load. For example, in a 20-mile wind a $4\frac{1}{2}$-pound load reduces the number of revolutions from 41 to 34, or 1.75 revolutions per brake pound. The power $= 2\pi\,RnL \div 33,000$, where $n =$ the number of revolutions of the wheel per minute, $L =$ the load on brake, in pounds, $R =$ the arm of brake (35 inches), $\pi = 3.1416$, and $33,000 =$ the number of foot-pounds per minute in a horsepower. We may write the power thus: $P = KnL$, and compute its value as follows, K being a constant equal to $2\pi R \div 33,000$:

$$
\begin{aligned}
P_0 &= K \times 41 && \text{(revolutions)} \times \ \ 0 \ \ (L) = \ \ \ 0 \\
P_1 &= K \times 33 && \text{(revolutions)} \times \ \ 4.5 \ (L) = 148.5K. \\
P_2 &= K \times 26 && \text{(revolutions)} \times \ \ 8.5 \ (L) = 221.0K. \\
P_3 &= K \times 22.5 && \text{(revolutions)} \times 10.5 \ (L) = 236.3K. \\
P_4 &= K \times 20.75 && \text{(revolutions)} \times 11.4 \ (L) = 238.6K. \\
P_5 &= K \times 19 && \text{(revolutions)} \times 12.5 \ (L) = 237.5K.
\end{aligned}
$$

P_4 is seen to be larger than any of the other values of P, and gives an approximate value of the power of the mill for that wind velocity (20 miles).

We may find a more accurate value in an easier way. Let x be the load for maximum horsepower in any given wind velocity. We have seen that 1 pound of load reduces the speed by 1.75 revolutions. Then we can write $P = K\,(41-1.75x)\,(x)$. For a maximum value of P we must differentiate P with respect to x, place the first differential coefficient 0, and solve for x. This value of x, according to calculus, makes the power a maximum. Differentiating, we have $P \div dx = 41-2\,(1.75x) = 0$. Solving for x we have $x = 11.7$ pounds. The corresponding value of the revolutions of the wheel per minute is $41-1.75 \times 11.7 = 20.5$ revolutions.

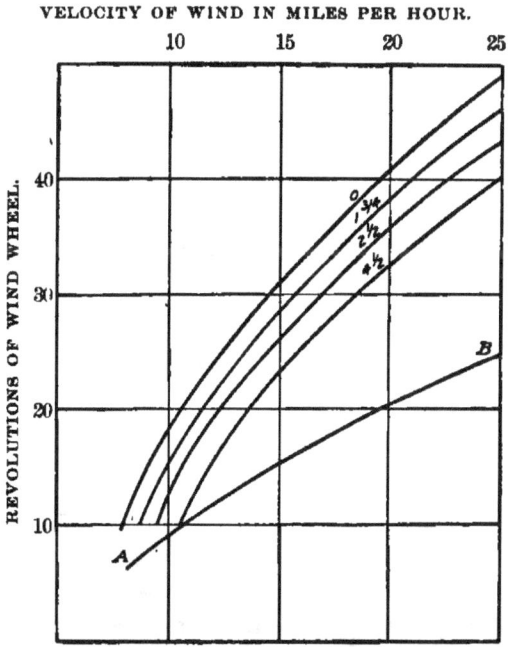

VELOCITY OF WIND IN MILES PER HOUR.

FIG. 62.—Diagram showing revolutions of wind wheel of mill No. 55—7.75-foot Jumbo. Curves marked 0, $1\frac{3}{4}$, $2\frac{1}{4}$, and $4\frac{1}{4}$ are for brake loads of 0, 1.75, 2.5, and 4.5 pounds, respectively; AB is for best load.

For a 25-mile wind we have $x = 49 \div 3.5 = 14$ pounds, and $n = 16.5$. In a similar way we get the load and revolutions for maximum power for other wind velocities. These and the corresponding horsepowers are as follows:

Values for the curve DK (maximum power) fig. 63.

Wind velocity (miles per hour).	Load in pounds.	Revolutions of wind wheel per minute.	Horsepower.
8	3.1	6.0	0.010
12	7.0	12.0	0.046
16	9.4	16.5	0.086
20	11.7	20.5	0.133
25	14.0	24.5	0.190

The curve *DK* is nearly a parabola. The revolutions per minute for the best load are seen to be very nearly equal to the wind velocity

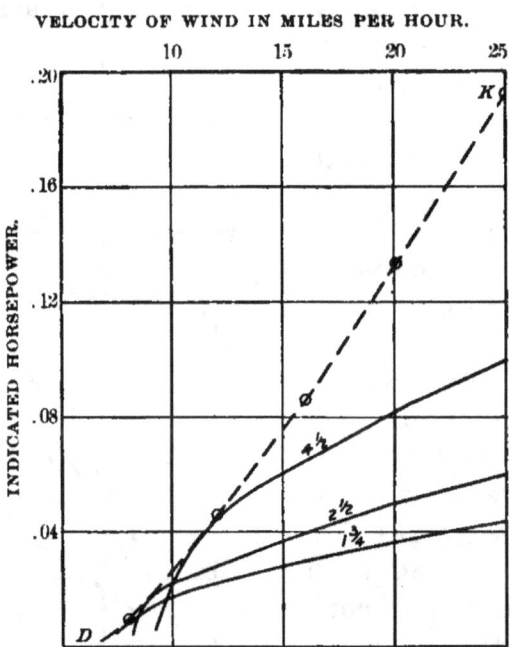

FIG. 63.—Diagram showing horsepower of mill No. 55—7.75-foot Jumbo. The curves show the horsepower for brake loads of 1.75, 2.5, and 4.5 pounds, respectively; dotted curve *DK* shows maximum power.

in miles. These are platted in fig. 62, giving the line *AB*, which is nearly straight.

RELATION BETWEEN WIND VELOCITY AND CIRCUMFERENCE VELOCITY OF WHEEL.

If we multiply the number of revolutions of the wind wheel for any wind velocity by 24.4 feet (the circumference of the wheel) and divide by the wind velocity, in feet per second, we have the following results for no brake load and for best load:

Table showing ratio between wind velocity and circumference velocity of wheel.

Load.	Wind velocity per hour.				
	8 miles.	12 miles.	16 miles.	20 miles.	25 miles.
No load	0.38	0.55	0.57	0.57	0.55
Best load	0.21	0.27	0.28	0.28	0.27

From this we see that the velocity of the circumference of the wheel is not more than 57 per cent of the velocity of the wind. For velocities above a certain amount it remains nearly constant for any load. It will be seen too that the speed of the wheel for best load is almost exactly half that for no brake load.

Test No. 2.—Sail area, $11\frac{1}{2}$ feet by 8 inches—each inner half sail removed. The results of this set of tests are as follows:

Results of tests of Jumbo mill No. 55, with 8 sails $11\frac{1}{2}$ feet by 8 inches.

Load.		Revolutions of wind wheel per minute at given wind velocities (per hour).					Horsepower at given wind velocities (per hour).				
		8 miles.	12 miles.	16 miles.	20 miles.	25 miles.	8 miles.	12 miles.	16 miles.	20 miles.	25 miles.
Pounds.	*Ft.-lbs.*										
0	0	9	22	30	39	47					
3.25	59.6	......	17	26	32	40	......	0.031	0.047	0.058	0.072

By comparing these results with those of test No. 1 it will be seen that the number of revolutions per minute for no load is from two to three times less when the half sails are used. The $3\frac{1}{2}$-pound load with half sails gives about the same speed as the $4\frac{1}{2}$-pound load with whole sails. The weight of the wheel is reduced about 40 per cent, which makes the reduction in speed less than it would be if the weight of the wheel remained constant. It will be seen, then, that for this size of wheel very little power is gained by the use of the inner 8-inch board of each sail. It is quite likely that sails 12 inches wide would give fully as much power as sails 16 inches wide.

Test No. 3.—The sail area was 4 sails, each $11\frac{1}{2}$ feet by 16 inches—every other full sail removed. The results of this set of tests were almost the same as those of test No. 1. There was no measurable reduction in the speed of the wheel when every other full sail was removed. The weight of the wheel was reduced about 40 per cent, and consequently the friction. The gain in pressure on the extra sail area is counterbalanced by the additional friction.

Test No. 4.—The sail area was 8 sails, each $11\frac{1}{2}$ feet by 16 inches—the same as for test No. 1. There was an inclined surface (shown in Pl. XVI, *B*) for concentrating air on sails and in a measure preventing an upward current from the front shield. The results of this set of tests are the same as those of No. 1. There was no measurable increase in the number of revolutions when the mill was loaded or unloaded, or when the incline was used or not used.

For this size of mill 4 sails each 12 inches wide give the maximum power. From our tests of other mills we should say that the sail width should increase directly as the diameter increases. For diameters of 12 feet or more it is likely that the addition of one or two more sails, say 5 for a 12-foot mill and 6 for a 16-foot mill, may increase the power over that for 4 sails.

In 1895 we made some measurements of the pressure of air on small curved surfaces,[1] from which we infer that if galvanized-iron sails curved to a radius about twice the width and with the concave surface to the wind were used the power of the mill would be increased about 15 per cent over that with the plane fans.

Putting the results of these tests of Jumbo mills in the most practical form, we have the following as the proper sail area and the probable horsepower of wooden Jumbo mills in a 16-mile wind when properly loaded, assuming the power to increase as the square of the diameter:

Table showing proper sail area and probable horsepower of Jumbo wooden mills in a 16-mile wind.

Diameter of wheel.	Number of sails.	Width of sails.	Length of sails.	Horsepower.
		Inches.	*Feet.*	
8 feet	4	12	12	0.09
12 feet	5	18	12	0.20
16 feet	6	24	12	0.36
20 feet	6	30	12	0.56

The formula used for computing the pressure on a series of plane surfaces moving in the direction of the velocity of the wind is—

$$P = F \frac{r}{g} (c - v)^2 \quad . \quad . \quad . \quad . \quad . \quad . \quad . \quad . \quad (a)$$

In this F is the area, in square feet, of the vane, r the heaviness of air at the observed temperature and barometric pressure, c the velocity of wind, and v the velocity of wind wheel, each in feet per second.

The heaviness of the air is found from the formula—

$$r = r_0 \frac{B}{B_0} \frac{T_0}{T} \quad . \quad . \quad . \quad . \quad . \quad . \quad . \quad (b)$$

In this T_0 is the absolute temperature in centigrade degrees.

From equation b we have $r = 0.08 \times \dfrac{29}{30} \times \dfrac{273}{301} = 0.07$ pound per square foot. From the table on page 123 it will be seen that for maximum power in a 16-mile wind $v = 0.28c$. Substituting in equation a we have, for the pressure on one sail—

$$P = 11.5 \times \frac{1}{32.2} \times 0.07 \, (1-0.27)^2 \, c^2 = 4.1 \text{ pounds.}$$

The arm of this pressure about the axis of the wheel is about 3.06 feet. Hence the moment of this force is $4.1 \times 3.6 = 14.76$ foot-pounds. This moment is equal to the moment of the brake load, and we have $14.76 = 35x$, and $x = 4.24$ pounds. The load actually carried on the brake, including friction, is about 8 pounds; hence nearly half of the working pressure comes from wind pressure on the approaching and receding sails, or only a little more than half the pressure comes from the sail which is at the highest position possible.

MATHEMATICAL DISCUSSION OF TESTS OF LITTLE GIANT MILL NO. 56.

This is a 4.67-foot mill made by Mr. C. Hunt, of Wichita, Kansas. These mills are made in sizes from 4 to 24 feet in diameter, to rest on a low tower or on a building. The largest one yet built is shown in Pl. XVI, A. It is used for grinding wheat. The Little Giant mill will be seen to resemble the Jumbo in that the wind wheel moves in the direction of the wind and not across it. It differs from the Jumbo, however, in having a vertical axis and many curved iron sails, instead of a horizontal axis and few plane wooden sails. The wind is prevented from striking the sails as they come around toward the wind by a shield which, when closed, covers about one-third of the circumference. The shield can move freely about the axis of the mill and has hinged to it a wing which can be held at right angles to the circumference. There is also a vane fastened to the shield, to aid in the government of the mill. When the wing of the shield is closed, the vane takes the direction of the wind and places the shield directly in front of the wind wheel, shutting off the wind from the wheel. When the wing is open, the pressure of the wind against it carries the shield around, admitting the wind to one-half of the wheel. By properly placing the vane and using the proper weight on the wing, the wind is admitted to a small or a large portion of the wheel, and thus the speed of the wheel is regulated. The mill receives the wind from all directions and regulates automatically.

Mill No. 56 has 24 curved iron sails, each 3 feet 10½ inches long and 6½ inches wide, set at an angle of 27° to the radius. The radius of curvature of the sails is 7½ inches. The vertical shaft of the wind wheel has a beveled cogwheel, which gears into another beveled cogwheel on a short horizontal shaft. The latter has a pitman for working a pump. The horizontal shaft on which the brake was placed was geared back 43÷13. The brake arm was 2 feet 10½ inches long. The number of revolutions per minute of the brake shaft was found for the four loads 0, 2, 4, and 6 pounds, respectively. The corresponding speed of the wind wheel is found by multiplying by 3¼. The results of the tests were as follows:

Results of tests of Little Giant mill No. 56.

Load.		Revolutions of brake shaft per minute at given wind velocities (per hour).					Horsepower at given wind velocities (per hour).				
		8 miles.	12 miles.	16 miles.	20 miles.	25 miles.	8 miles.	12 miles.	16 miles.	20 miles.	25 miles.
Pounds.	Ft.-lbs.										
0	0	10	19	25	31	38					
2	35.7	6	16	21	27	34	0.007	0.017	0.023	0.030	0.040
4	71.4	......	11	18	23	30	......	0.024	0.039	0.050	0.065
6	107.1	......	6	14	19	26	......	0.019	0.045	0.062	0.084

Referring to fig. 64, the mill will be seen to start for no load in a light wind—about 5 miles an hour. More than half of each speed curve is a nearly straight line. This is due to failure to govern, the mill being held wide open all of the time. It will be seen that each pound of load reduces the speed about two revolutions per minute for about 0.67 of each curve. Fig. 65 shows the brake horsepower for three loads—2, 4, and 6 pounds, respectively. The curve DK, to which these load curves are tangent, shows the maximum power of the mill. This curve passes through the points y (horsepower)=0 when x (wind velocity)=5, and y=0.024 when x=12. Hence, if we assume this curve DK to be a parabola, we have for its equation—

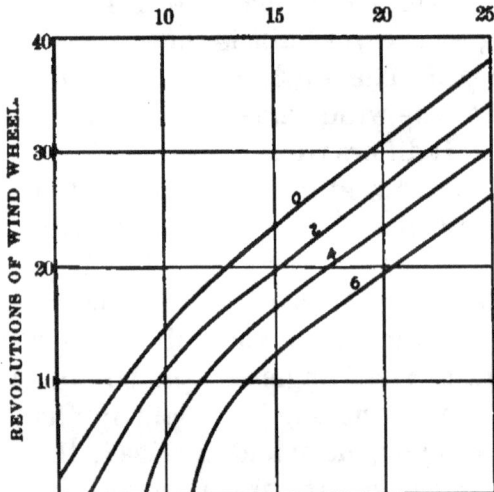

VELOCITY OF WIND IN MILES PER HOUR.

FIG. 64.—Diagram showing revolutions of wind wheel of mill No. 56—4.67-foot Little Giant. The curves marked 0, 2, 4, and 6 are for brake loads of 0, 2, 4, and 6 pounds, respectively.

$$x^2 = 496y + 25.$$

The horsepower from this equation and from fig. 65 is as follows, y' being taken from fig. 65:

Horsepower of mill No. 56 at given wind velocities.

x	y	y'
5	0.000	0.000
8	0.008	0.007
12	0.024	0.024
16	0.046	0.045
20	0.075	0.065

It will be seen that the horsepower of this mill does not increase as fast as the square of the wind velocity. From fig. 65 it will be seen that the 2-pound curve is tangent at 8 miles, the 4-pound curve at 12 miles, and the 6-pound curve at 16 miles an hour. Hence the load for maximum power increases about as the first power of the wind velocity. The speed of the brake shaft for these loads is 6, 11, and 14 revolutions per minute, respectively. Hence the speed of the wind wheel does not increase as fast as the first power of the wind velocity. The ratios of circumference velocity of wind wheel to wind velocity for no load and for maximum load for five wind velocities are as follows:

Table showing relation of circumference velocity of wind wheel to wind velocity.

Wind velocity per hour.	No load.			Maximum load.			Ratio of revolutions at maximum load to revolutions at no load.
	Revolutions of brake shaft per minute.	Circumference velocity in miles per hour.	Ratio of circumference velocity to wind velocity.	Revolutions of brake shaft per minute.	Circumference velocity in miles per hour	Ratio of circumference velocity to wind velocity.	
8 miles	10	5.2	0.65	6	3.1	0.39	0.60
12 miles	19	9.9	0.74	11	5.7	0.47	0.58
16 miles	25	13.0	0.81	14	7.3	0.46	0.56
20 miles	31	16.1	0.80	16	8.3	0.42	0.52
25 miles	38	19.8	0.79	....	....	....	....

From this we see that for no load the greatest circumference velocity is only 81 per cent of the wind velocity. For best load the ratio is 47 per cent. Here is the great disadvantage of these wheels, which move in the direction of the wind instead of across it—they move too slowly. The ratio of circumference velocity to wind velocity in an Aermotor is 1.75. This is 2.16 times the greatest corresponding circumference velocity of the Little Giant mill.

COMPARISON OF LITTLE GIANT AND JUMBO MILLS.

From the following table it will be seen that the circumference velocity of the Jumbo is only from 0.55 to 0.70 of that of the Little Giant for best load. The horsepower of the Jumbo is about 1.9 times that of the Little Giant; the ratio of the Jumbo diameters is 1.67;

hence the ratio of the power is a little greater than that of the diameters. It must be remembered, however, that the Jumbo is about three times the length of the Little Giant. For the same sail lengths the latter does 1.57 more work than the Jumbo.

Comparative data of Little Giant and Jumbo mills.

Mill.	Maximum horsepower at given wind velocities (per hour).					Circumference velocity for maximum load at given wind velocities (per hour).				
	8 miles.	12 miles.	16 miles.	20 miles.	25 miles.	8 miles.	12 miles.	16 miles.	20 miles.	25 miles.
7½-foot Jumbo	0.010	0.046	0.086	0.133	0.190	1.7	3.40	4.70	5.80	------
4½-foot Little Giant	0.007	0.024	0.045	0.066	0.100	3.1	5.70	7.30	8.30	------
Ratio	1.4	1.9	1.9	2.0	1.9	0.55	0.60	0.64	0.70	------

Taking into account the difference in the diameters of the wind wheels of these mills, we may say that the Little Giant will furnish about 2.5 more power than the Jumbo for the same diameter and length of sail. The Jumbo requires a 7-mile wind to start it with no load; the Little Giant will start with no load in less than a 5-mile wind. The Jumbo has no means of governing; the Little Giant governs easily and completely. The Jumbo gets the full pressure of the wind when it comes from two directions only; the Little Giant works equally well with the wind from any direction. The Little Giant is less likely to be injured in a windstorm than the Jumbo. The first cost of the Jumbo is somewhat less than that of the Little Giant. A 5-foot Little Giant with stub tower can be bought for about $15.

COMPARISON OF LITTLE GIANT WITH 8-FOOT AERMOTOR.

The efficiency of the pump and well of Aermotor No. 5 (for description and results of tests, see pages 33 and 34, Part I) is probably about 60 per cent. For the speeds and horsepowers of these mills we have the following:

Comparative data of Little Giant and 8-foot Aermotor.

Mill.	Load per revolution of wind wheel.	Sail area.	Revolutions of wind wheel per minute at given wind velocities (per hour).				Horsepower at given wind velocities (per hour).			
	Ft.-lbs.	Sq.ft.	12 miles.	16 miles.	20 miles.	25 miles.	12 miles.	16 miles.	20 miles.	25 miles.
8-foot Aermotor (No. 5)	45	34	62	84	99	115	0.084	0.115	0.138	0.158
4½-foot Little Giant (No. 56)	24	54	34	55	71	92	0.024	0.039	0.050	0.065
Ratio	1.96	0.63	1.82	1.53	1.40	1.25	3.50	3.00	2.76	2.43

It will be seen that the sail area of the Aermotor is only 0.63 that of the Little Giant, that the wind wheel of the former makes from 1.25 to 1.82 more revolutions per minute than the latter, and that the power of the former is from 3.5 to 2.4 times greater than the power of the latter. While the sail area of the Aermotor is only 0.63 that of the Little Giant, the wind area of the former is much greater than that of the latter. The wind can not enter the wheel of the Little Giant over an area greater than the radius multiplied by the length of the sail, or 9.1 square feet. In the Aermotor the wind enters the wheel over an area equal to the difference between the areas of the two circles, one having a diameter of 8 feet, the other having a diameter of 3 feet. This wind area is 43.2 square feet. Hence the wind area of the Aermotor is $43.2 \div 9.1 =$ 4.75 times that of the Little Giant. Here is the great advantage that the Aermotor has over the Little Giant—it has only 0.63 of the sail area, and hence cost less for sails, and 4.75 times more air strikes its sails than strikes those of the Little Giant. It will be shown later that in the Little Giant the air acts on its sails while passing out of the wind wheel as well as while passing into the wheel, and thus the power for the same wind area is greater in that mill than in the Aermotor.

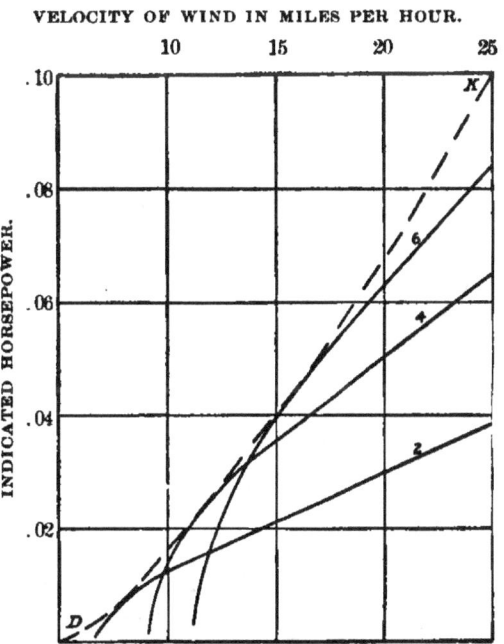

FIG. 65.—Diagram showing horsepower of mill No. 56—4.67-foot Little Giant. The curves marked 2, 4, and 6 show horsepower for brake loads of 2, 4, and 6 pounds, respectively; dotted curve DK shows maximum power.

Fig. 66 is a diagram showing the action of the wind on the sails of the Little Giant mill. AH, BK, etc., are the curved sails. The cord AH makes an angle of 27° with the radius AP. CD is the shield, with the wing DE open. Let Aa represent the direction and magnitude of the wind with respect to the earth. We have seen that for maximum load the ratio of the circumference velocity of the wheel to the wind velocity is 0.47; hence drawing Ac tangent to the circumference Aa and equal to 0.47 of Aa, and completing the parallelogram on them, we have Ab representing the direction of the wind with respect to the moving sail. If we assume this to be the velocity of the air over the sail (it is somewhat less than this, since Ab is not quite tangent to the sail at entrance), we can construct the path of a particle of air with respect to the earth. The points 1, 2, and 3 are on this path, and $L3$ gives the approximate direction of the particle of air through the wheel. At L we combine this velocity with the inner circumference velocity of the sail, giving the velocity Lp with respect to the moving sail. This is seen to make a large angle with the tangent at entrance, reducing its magnitude somewhat. $L4$ is approximately the absolute path of the particle over the sail, and 4-5 is the direction of the particle at exit. $B6$ represents the absolute path of a particle of air as it moves over the sail BK.   This particle of air passes through the

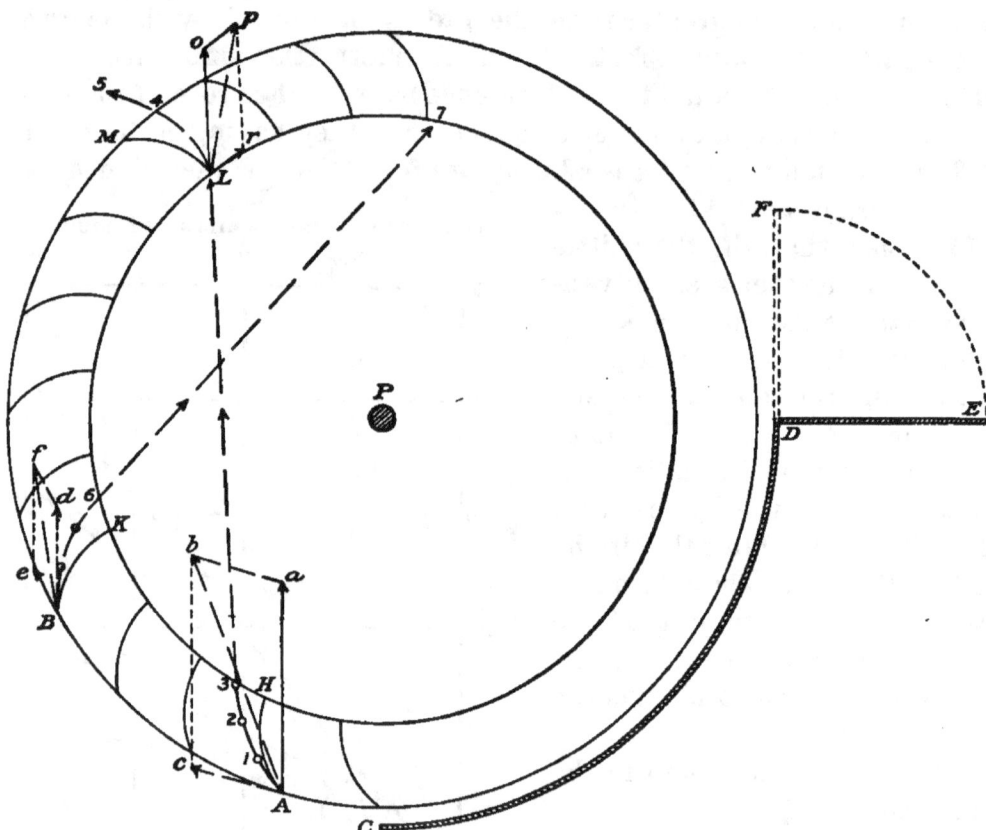

FIG. 66.—Diagram showing action of wind on sails of mill No. 56—4.67-foot Little Giant.

wheel in the direction 6-7.   The path 6-7 of this particle crosses the path of the particle from 3; hence there is interference inside the wheel, which prevents our tracing with accuracy the path of a particle out of the wheel.   It is evident, however, that after passing into the wheel the air strikes the concave side of the sails on the opposite side, and aids in pushing the wheel around, so all the work is not done by the sails on the side where the air enters.

INDICATED AND TRUE VELOCITIES.

Thus far all of our results for speed and power of windmills are given in terms of indicated velocities, that is, velocities as read from the Robinson cup anemometer. It is necessary, or at least desirable, to examine these to see whether they agree with true velocities or distances actually passed over by the wind per hour. Fig. 67 shows the Robinson cup anemometer as used by the United States Weather Bureau. This instrument was invented in 1846 by Dr. T. R. Robinson, of Armagh, Ireland, and is now used by several meteorological bureaus for the measurement of wind velocity. It gives a continuous record of wind movement and requires no device, such as a vane, to give it the proper direction with respect to the wind. It is made so that each 50 revolutions of the cups can be read on the dial, and there is an electrical device for recording each 250 or each 500 revolutions of the cups.

Referring to fig. 68, let $A =$ the upper and $B =$ the lower cup of a Robinson anemometer rotating about the axis, let c equal the velocity of the wind, and v the velocity of the cup center, each in feet per second; $x =$ the ratio of the velocity of the wind to that of the cup center, $P_1 =$ the pressure on the concave surface, and $P_2 =$ the

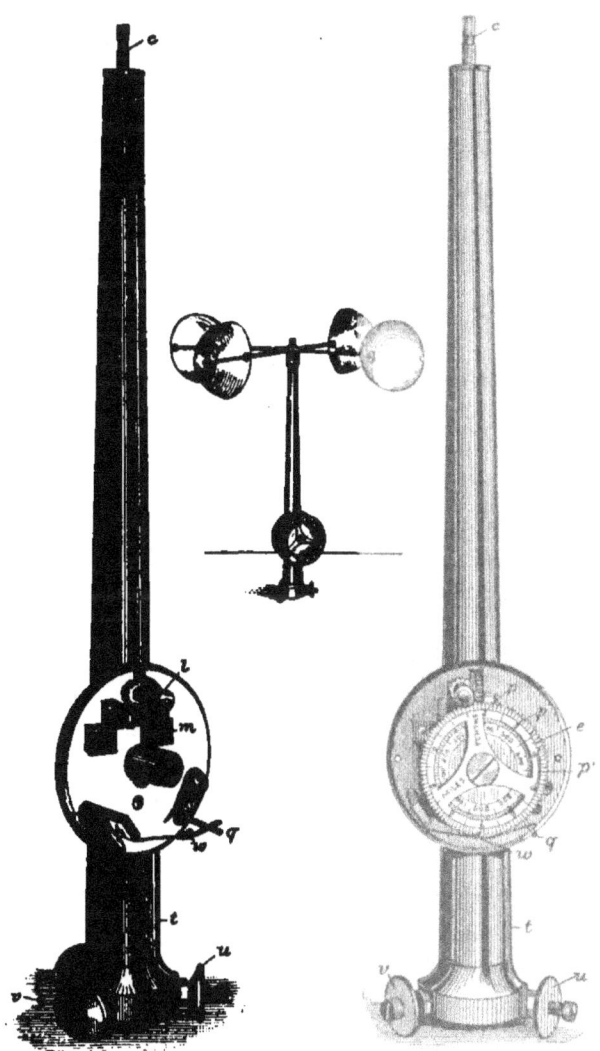

Fig. 67.—Anemometer and cups. c, spindle which forms the axis of revolution of the cups; m, 50-tooth wheel which engages an endless screw on the end of the spindle; l, small toothed wheel which engages an endless screw on the axis of the wheel m; e, pair of dial wheels which are moved by the wheel l; p, one of ten contact pins to aid in closing the electric circuit at the end of each mile of wind: p', two of these pins connected, forming the tenth mill pin: w, contact spring; q, a contact point at the end of contact spring; t, small insulated tube connecting q with the insulated binding post u and with the second binding post v.

pressure on the convex surface of the cup. Dr. Robinson found from sixteen experiments with stationary cups exposed to wind of several velocities that for all velocities the pressure when the concave surface of the cup is toward the wind is about four times that when the convex surface is presented to the wind. The pressure on the moving cup A is $P_1 = K_1 Fr (c-v)^2 \div 2g,$[1] and the pressure on the cup B is $P_2 = K_2 Fr (c+v)^2 \div 2g;$ F being the area of the cup, K_1 and K_2 being constants the ratio of which, as found by Robinson, is 4, r the heaviness of air, and $c+v$ and $c-v$ the relative velocities of the cups. Neglecting friction in the anemometer, inertia of cups and arms, and the influence of two of the cups until they are near the position shown in fig. 68, we see that for uniform velocity P_1 must equal P_2; if P_1 is greater than P_2, v will increase; if P_1 is less than P_2, v will decrease. We have, therefore, for uniform velocity $K_1 Fr (c-v)^2 \div 2g = K_2 Fr (v+c)^2 \div 2g$ or $4 (c-v)^2 = (c+v)^2$. This equation can be put in the form $4 (x-1)^2 = (x+1)^2$. Solving we have $x = 3$; that is, the velocity of the wind is three times that of the cup centers. For an anemometer having arms 6.72 inches long, the distance passed over by a cup center in 500 revolutions is

$$d = \frac{2 \times 22}{7} \times \frac{6.72}{12} \times 3 \times 500 = 5,280 \text{ feet, or 1 mile.}$$

Dr. Robinson believed that this ratio of wind velocity to that of

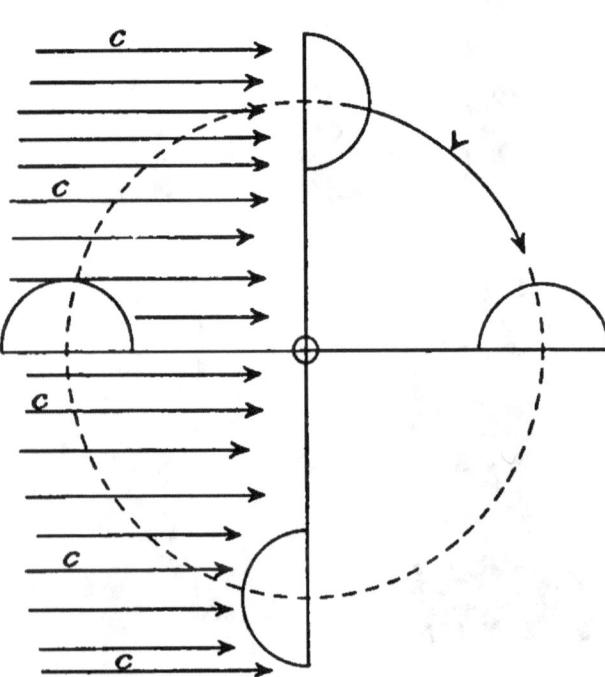

FIG. 68.—Diagrammatic section of anemometer cups.

cup center was true for all velocities, and consequently the makers of the instrument have marked the dial in miles. We shall see, however, from the rating of the one that we have used in these windmill tests that this ratio is not a constant. It will be seen that for wind of uniform velocity this ratio is a variable which has the value 3 for 9 miles an hour, is greater than 3 for less velocities, and is less than 3 for greater velocities than 9 miles an hour; in other words, to get true velocities we must add a correction below 9 miles an hour and subtract a correction above that velocity. Friction and inertia were neglected in deriving the foregoing value of this ratio. The former has little influence in an instrument kept in good condition, but in a poorly kept instrument it may have a large influence for low velocities. This ratio has been found to be 8 to 10 with much friction. The inertia of the arms and cups has a marked influence on this ratio, especially for ordinary gusty wind. As the

---

[1] Church's Mechanics of Engineering, p. 815.

gustiness of the wind increases the correction to be subtracted to get the true velocity increases also.

The relation between the indicated and true velocity of an anemometer is found by moving the anemometer in still air at different velocities, and noting the distance passed over, also the readings of the instrument. It is seldom that the air out of doors is still for any considerable length of time, so that this comparison is usually made within an inclosure, the anemometer being carried around in a circle. The radius of the whirler should be as long as possible, and made so as to affect the circulation of air as little as possible, and to reduce the effect of the centrifugal force.

The whirling machine that we have used for rating the anemometer is the property of the United States Weather Bureau. It consists essentially of an arm 28 feet long and 8 feet above the ground, on the end of which the anemometer is carried at an elevation of 2 feet above the arm. This arm is counterweighted and is stiffened by the rods. It is clamped to a vertical shaft, which carries a cogwheel near its lower end. A cogwheel on a horizontal shaft engages the large cogwheel and gives rotation to the arm. For low velocities the power was applied through a crank on the horizontal shaft, and for higher velocities by a crank on a second shaft, the latter working the first shaft by means of two sprocket wheels and a chain. The machine was set up out of doors, in a sheltered place away from any building, and was used on several nights when there was scarcely any wind. It was made to rotate about half the time in the positive direction and the other half in the negative direction.

A new Robinson anemometer was used with which to compare results obtained for the one used in our windmill tests. The results for these instruments agree quite closely. The results obtained on December 29, when there was no perceptible wind, are as follows:

Table showing relation between indicated and true velocity of anemometer.

Indicated velocity, in miles per hour.	Revolutions of long arm in ⅛ mile of wind.	True velocity, in miles per hour.	Correction, in miles.	Correction, in miles, for gusty wind.
6	15.5	6.20	+0.20	0
8	15.1	8.05	+0.05	−0.2
11	14.7	10.78	−0.22	−0.6
15	14.2	14.20	−0.80	−1.2
20	13.8	18.40	−1.60	−2.2
25	13.53	22.55	−2.45	−3.2
30	13.3	26.60	−3.40	−4.3

It will be seen that the indicated velocity is less than the true velocity for velocities less than 9 miles an hour; above 9 miles the true velocity is less than the indicated velocity. In other words, the correction is added below 9 miles an hour and subtracted above that velocity. It will be seen that up to 11 miles an hour these anemometer readings differ very little from the true readings, but for higher velocities the correction becomes quite large. The last column of the foregoing table gives the corrections to be applied to the indicated velocities of the Weather Bureau Robinson anemometer for gusty wind.[1] The motion of ordinary moving air, when studied with a very light anemometer recording each revolution, is found to vary suddenly by large amounts. The rate of motion changes 20 or 30 miles an hour in a few seconds.[2] The record of the standard Robinson anemometer, recording miles or half miles, does not show these sudden changes, but gives an average velocity for the wind. Its

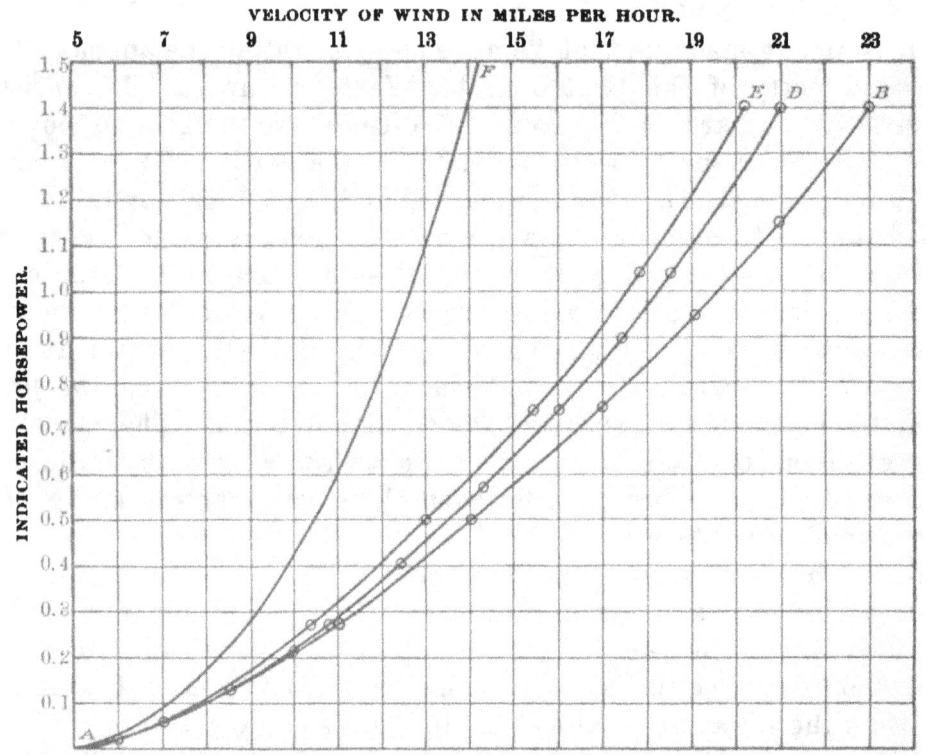

FIG. 69.—Diagram showing horsepower of mill No. 27—12-foot Aermotor. Curve *AB* shows power for indicated velocity, assuming wind to be not gusty; *AD* shows power for true velocity; *AE* shows power in average gusty wind; *AF* shows power from H. P. $= \dfrac{x^3 - 140.6}{2,086}$.

weight and consequent inertia cause it to continue its rotation for a time after the impulse is passed, and when the next impulse strikes the cups their weight will not allow them to take the velocity of the impulse. The less the weight the more nearly will the velocity of the cups be that of each gust. The effect is that the cups revolve faster in a gusty wind of, say, 20 miles an hour, than in wind of the same velocity but not gusty. Two difficulties arise in dealing with gusty wind: (1) The gustiness of any wind varies from time to time, and the anemometer gives no indication of it; (2) we have no means of producing artificial gusty wind in which to rate anemometers. If the anemometer is rated in natural wind, then a correction

---

[1] Anemometry, by C. F. Marvin. Washington, 1893.
[2] The Internal Work of the Wind, by S. P. Langley. Washington, 1893.

must be applied, but its amount is uncertain. The velocity of gusty wind can not, therefore, be measured with certainty. This, however, does not lessen the value of the Robinson anemometer as an instrument for measuring wind velocity, nor introduce an error in our results. For a given velocity and gustiness the anemometer will always give the same reading, and the same gusty wind which strikes the anemometer strikes the windmill directly behind it. The difficulty arises when we try to compare results in which the Robinson anemometer was used in measuring velocity with those in which some other form of anemometer was used. The corrections given in the last column of the table were found by Prof. C. F. Marvin, of the United States Weather Bureau, and are for ordinary gusty wind.

The Robinson anemometer is now so generally used to measure wind velocity that it is better to express the speeds and powers of windmills in terms of these than of any other. If results are desired in terms of true velocities, they can be found approximately by means of the table of corrections given on page 133. Fig. 69 shows the power of mill No. 27 expressed in terms of three kinds of velocity: AB shows the power assuming the wind to be not gusty; AD shows the power for true velocity, that is, after applying the corrections in the fourth column of the table; AE shows the power in average gusty wind, that is, after applying the corrections in the last column of the table; AF would give its power if the power increased as the cube of the indicated velocity.

COMPARISON OF WRITER'S EXPERIMENTS WITH THOSE OF OTHERS.

Comparison with Smeaton's experiments.—In this comparison we use indicated velocities and results for best mills. Smeaton's results are given on pages 15 and 16, Part I. It will be remembered that his wheels were 3.5 feet in diameter, moved against still air in a circle of 5.5 feet radius, and that his wind velocities varied from 3 to 6 miles an hour.

Smeaton found (maxim 1, page 15, Part I) that the velocity of a windmill sail, whether loaded, so as to produce maximum power, or unloaded, is nearly as the velocity of the wind. We have found that the velocity of wind wheel when loaded increases nearly as the wind velocity, but when it is unloaded it increases as the square root of the wind velocity.

Smeaton found (maxim 3) that the maximum power increases somewhat less rapidly than as the cube of the wind velocity. We have found that the maximum power increases as the square of the wind velocity—for true velocities somewhat faster than as the square of the wind velocity.

Smeaton found (last part of maxim 5) that the power for a constant load increases as the first power of the wind velocity. Our experiments show that for a constant load the power increases as the square root of the wind velocity.

Smeaton found (maxim 6) that the circumference velocities of similarly made mills of different diameters vary inversely as the diameters. We have found this to be true.

Smeaton found (maxim 8) that the maximum power for similarly made mills increases as the squares of the diameters. We have found that it increases about as the 1.25 power of the diameters.

Smeaton found (maxim 9) that the circumference velocity of the Dutch sail, whether loaded or unloaded, is considerably greater than the wind velocity. We find that it is nearly equal to the wind velocity for loaded sail, and about 1.75 times the wind velocity for the unloaded sail.

Comparison with Coulomb's experiments.—In this comparison indicated velocities are used. Coulomb's observations (see page 16, Part I) were made on a Dutch mill (fig. 1, Part I) having a wind wheel 70 feet in diameter. He found that at a wind velocity of about 15 miles an hour the wind wheel was making 13 revolutions per minute and yielding about 7 horsepower of useful work. Comparing this with the results for the 16-foot Aermotor for maximum load, we have the following:

Comparison of results of tests of Coulomb's 70-foot mill with writer's 16-foot Aermotor.

Mill.	Revolutions of wind wheel per minute.	Circumference velocity in feet per minute.	Horsepower.
70-foot mill	13	2,860	7.0
16-foot Aermotor	28	1,408	0.8

The circumference velocity of the large mill is more than twice that of the smaller one. The ratio of the powers is $7 \div 0.8 = 8.8$. The ratio of the diameters is $70 \div 16 = 4.4$. The ratio of the powers is about twice that of the diameters. The ratio of the squares of the diameters is 19.14.

It is very likely that the wind velocity as found by Coulomb is too small; the very large circumference velocity of his wheel indicates this. It is probable that the wind velocity during his observations was about 20 miles an hour instead of 15. In a 20-mile wind the horsepower of the 16-foot mill is about twice that in a 15-mile wind, and the speed 34 revolutions per minute against 28. The ratio of the horsepowers would then be about as the diameters of the wind wheels.

Comparison with Griffiths's experiment.—The performance of a pumping windmill depends on so many factors, most of which may affect the result to a large degree, that it is doubtful whether it is worth while to make a comparison where the conditions differ much. There is the pump efficiency alone, other conditions being the same, which may make the horsepower of one four or more times that of the other. For example: Both mills have a brake power of 1 horsepower; one mill works a pump which has an efficiency, under present conditions, of 20 per cent, its useful horsepower being 0.2; the other mill operates a pump which has an efficiency of 80 per cent, its horsepower being 0.8, or four times that of the other. We have seen that for wind velocities above a certain amount the power increases nearly as the load, so that by doubling the load for the higher velocities the power is doubled. The gearing and means of governing affect the power in a somewhat less degree. The way in which the wind velocity is measured may affect the recorded power and speed. If the wind wheel, or tower, or any other obstacle obstructs the free flow of the air to the anemometer, the recorded velocity will not be as great as it should be. If the anemometer is placed on the platform, it will give a less velocity than if held some distance in front of the wheel and at the height of the axis. The temperature and barometric pressure affect the power.

In Mr. Griffiths's data (see pages 17 and 18, Part I) the load factor is known, but the wind velocities he gives are small, and for small velocities it is difficult to compute the effect on the power of differences in load. In fact, most of his velocities are less than are required to start irrigating mills. Only four of his mills, viz, Nos. 1, 2, 5, and 6, are comparable with mills that we have tested. The mills with which we have compared them are Nos. 36, 38, and 47—very lightly loaded mills with low pump efficiency.

Comparison of results of Griffiths's tests with those of writer's tests.

Mill.	Outer diameter of sail.	Load per stroke of pump.	Wind velocity.	Strokes of pump per minute.	Horsepower.
	Feet.	*Ft.-lbs.*	*Miles.*		
Griffiths's No. 1	22.3	480.0	7.0	6.8	0.098
Writer's No. 36	22.5	248.0	7.0	5.0	0.038
Griffiths's No. 2	11.5	29.2	5.8	13.0	0.011
Writer's No. 38	10.0	21.0	5.8	9.0	0.006
Griffiths's No. 5	10.2	51.0	8.5	20.5	0.028
Writer's No. 38	10.0	21.0	8.5	21.0	0.014
Griffiths's No. 6	9.8	30.7	6.0	12.5	0.012
Writer's No. 47	10.0	37.0	6.0	7.0	0.008

It will be seen from this table that the horsepowers of the mills tested by Mr. Griffiths are greater than the horsepowers of the mills we have tested. The speeds of the wheels are also greater, except in Nos. 5 and 38. We are of the opinion that a part of the difference in results is due to difference in the method of measuring the wind velocity. The wind velocity found by Mr. Griffiths is less than we have found it.

Comparison with King's experiments.—Professor King's measurements of the brake horsepower of a 16-foot Aermotor are given on page 20, Part I.   These results are plotted in fig. 70, giving the curve *AB.*   The curve of maximum power for the 16-foot Aermotor No. 44 is redrawn in this figure as curve *CD.*   It will be seen that these curves are nearly parallel to 16 or 17 miles an hour, and then diverge rapidly. Believing that much of this difference is due to the wind wheel of Professor King's mill interfering with the anemometer, we have experimented with two anemometers, one located 29 feet directly south of the wind wheel of mill No. 44, the other located 27 feet directly north of it, and each recording, side by side, the wind velocity on an electric register.   The hourly wind velocity for fourteen consecutive hours, during the first ten of which the mill was working and during the last four of which it was out of the wind, is given in the following table:

Table showing wind velocities during writer's experiments with anemometers on mill No. 44.

Hour.	Direction of wind.	Wind velocities.		Ratio.
		Front anemometer.	Rear anemometer.	
		Miles.	*Miles.*	*Per cent.*
First	SE	15.0	10.2	0.68
Second	do	18.0	13.4	0.74
Third	do	18.0	13.5	0.75
Fourth	do	19.5	12.8	0.65
Fifth	SW	18.0	13.5	0.74
Sixth	do	16.3	12.6	0.77
Seventh	S	15.2	10.0	0.65
Eighth	SE	12.7	9.6	0.76
Ninth	do	10.5	7.5	0.71
Tenth	do	7.7	5.5	0.72
Eleventh	do	7.2	7.2	1.00
Twelfth	do	8.6	8.6	1.00
Thirteenth	do	10.5	10.6	1.00
Fourteenth	do	10.3	10.4	1.00

The direction of the wind is the mean for each hour, as shown by an anemoscope.   When the wind was from the southeast it occasionally came for a short time almost directly from the east; and, again, when from the southwest it occasionally came for a time from the west.

It will be seen that the hourly velocity, 27 feet, behind the running wheel was only 65 to 77 per cent of that in front of the wheel.   As soon as the wheel was turned out of the wind the two anemometers recorded nearly the same velocity.

Referring now to the direction of the wind when Professor King's windmill tests were made, and remembering that his anemometer was 40 feet directly east of the moving wheel, it will be seen that nearly all of the tests were made when the wind came from the northwest or the southwest, and that consequently the moving wheel must have interfered with the proper working of the anemometer, causing it to record a less velocity than actually existed, and making the horsepower greater than if the anemometer had been in front of the wheel. In fig. 70 it will be seen that an increase of from 5 to 25 per cent in Professor King's wind velocities would move his curve over to the right of our curve. These curves are found in very different ways—Professor King's from 26 single observations, ours from more than 150 observations. Instead of finding points on this curve when the proper load is unknown, we have found speed and power curves for constant loads, and from these drawn the curve of maximum power. None of the mills that we have tested have given a power curve like AB in fig. 70. It will be noticed too that the curve CD is quite like the corresponding one for the 12-foot Aermotor. This we should expect, since the mills are similar in construction.

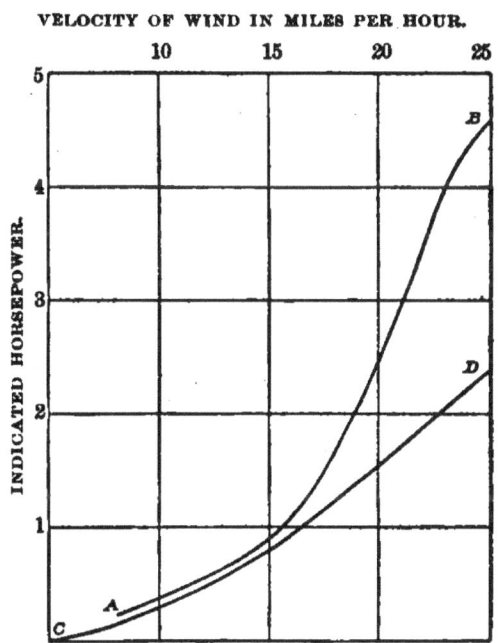

FIG. 70.—Diagram showing horsepower of two 16-foot Aermotors. Curve AB shows brake horsepower of Professor King's 16-foot Aermotor; CD shows maximum power of writer's 16-foot Aermotor No. 44.

Comparison with Perry's experiments.—Indicated velocities are used in this comparison. Some of Mr. Perry's results are given on pages 20 and 21, Part I. His tests were made on wheels 5 feet in diameter, carried against still air in a circle 14 feet in diameter, and his wind velocities were not greater than about 11 miles an hour. Mr. Perry states that his results agree with those of Smeaton. A comparison of our results with Smeaton's has been given on pages 135 and 136.

We will compare in detail the working of two of Mr. Perry's wheels, viz, Nos. 44 and 48, with that of our 12-foot Aermotor No. 27.

Mr. Perry's wheel No. 44 is somewhat like the wind wheel of our 12-foot Aermotor No. 27. It has 12 curved wooden sails, each 18 by 12.3 by 5.8 inches, having a weather angle at the inner end of sail of 30° and an angle of 25° at the outer end. The air is more obstructed in its passage through this wheel than through the wheel of No. 27. The following are some of the results for these mills in an 8.5-mile wind, the only velocity which Mr. Perry gives for his mill.

Comparison of results of tests of Perry's 5-foot mill No. 44 with writer's tests of 12-foot Aermotor No. 27.

Mill.	Area of sail.	Wind velocity.	Load per revolution.	Maximum load.		Horse-power.	No load.	
				Revolutions per minute.	Circumference velocity.		Revolutions per minute.	Circumference velocity.
	Sq. ft.	*Miles.*	*Ft.-lbs.*					
Perry's mill No. 44..	13.6	8.5	11.3	44.4	8.6	0.016	84.3	22.1
12-foot Aermotor No. 27............	73.0	8.5	22.2	19.0	11.9	0.13	32.5	20.4
Ratios..........	5.3	....	19.6	....	1.38	8.12	....	0.92

It will be seen that the load, in foot-pounds, per revolution of wind wheel is 19.6 times greater and the circumference velocity for maximum load is 1.38 times greater for No. 27 than for the 5-foot wheel, but that the circumference velocity for no load is a little less for No. 27 than for the other mill. The power of the 12-foot Aermotor is more than 8 times that of the 5-foot mill, and its sail area is 5.3 times greater.

We will next compare the 12-foot Aermotor No. 27 with Mr. Perry's 5-foot mill No. 48, which gave the greatest power of the 61 wheels tested by him. It had six curved pasteboard sails, each 19 by 23.7 by 10.9 inches, set at a weather angle of 35° at the inner end of the sail, and at an angle of 25° at the outer end. All obstructions to the free flow of air over the back of the sails were removed. The following are some of the results for these mills at a wind velocity of 11 miles an hour:

Comparison of results of tests of Perry's 5-foot mill No. 48 with writer's tests of 12-foot Aermotor No. 27.

Mill.	Area of sail.	Wind velocity.	Load per revolution.	Maximum load.		Horse-power.	No load.	
				Revolutions per minute.	Circumference velocity.		Revolutions per minute.	Circumference velocity.
	Sq. ft.	*Miles.*	*Ft.-lbs.*					
Perry's mill No. 48 .	14.2	11	23.3	66.8	17.4	0.047	142	37.2
12-foot Aermotor No. 27............	73.0	11	333	28.5	17.9	0.288	45	28.3
Ratios	5.2	....	14.3	....	1.03	5.96	....	0.76

It will be seen that the circumference velocities of these wheels for maximum power are nearly equal, but for no load the circumference velocity of the 5-foot mill is about 25 per cent greater than that of the 12-foot Aermotor. The power of the latter mill is nearly six times that of the 5-foot mill. There are more air obstructions in the wheel of the 12-foot mill than in that of the 5-foot mill, so that the difference in the power would be greater for equal air obstructions. For a corresponding amount of obstruction the ratio of power would probably be 6.5 to 7. This ratio is greater even than $12 \div 5^2 = 5.76$, the ratio of the squares of the diameters.

From this comparison of the results of our tests with those of Smeaton and Perry it will be seen that the power of a natural moving air of a given measured velocity is greater than the resistance of the air to a wheel carried around in a circle. Some of the laws (see page 114) which have been found to govern wheels moved against still air— notably that the power increases as the cube of the wind velocity—are not applicable to windmills in moving air.

ECONOMIC CONSIDERATIONS.

The power of windmills has been computed from tests on model windmills, in artificial air of low velocity, assuming, first, that the power increases as the cube of the wind velocity, and, second, that the power increases as the square of the diameter. Our tests of windmills recorded in the preceding pages show that the power does not increase much faster than as the square of the wind velocity, and about as 1.25 times the power of the diameter of the wind wheel. We believe that to these two false assumptions is due the exaggerated power of windmills claimed by windmill makers and others interested. A good 12-foot steel mill should furnish 1 horsepower in a 20-mile wind (indicated) and 1.4 horsepower in a 25-mile wind. This is the smallest amount of power that will do any considerable amount of useful work. A 16-foot mill will furnish 1.5 horsepower in a 20-mile wind (indicated) and 2.3 horsepower in a 25-mile wind.

A 12-foot steel mill and a 50-foot steel tower as commonly made weigh about 2,000 pounds. A 16-foot steel mill and a 50-foot steel tower weigh about 4,250 pounds. The 16-foot outfit weighs more than twice that of the 12 foot, and its power is only 1.5 that of the latter. In addition, the 12-foot mill will govern more easily and is less likely to be injured in a storm than the 16-foot mill. In most cases, therefore, it is better to use two 12-foot mills than one 16-foot mill.

The economic value of a windmill depends on its first cost, on the cost of repairs, and on its power. Most of the effort put forth at the present time to improve windmills is directed toward reducing the first cost. Competition is so strong that the cost must be kept low, and this is often accomplished at the sacrifice of the other two factors— cost of repairs and power. The pumping mills and their towers are, as a rule, too light and lacking in stiffness. It is said that in some parts of the West wooden mills are coming into use again, on account of the lightness and poor quality of the steel mills. This, however, is a fault of the making, not of the material. The wooden tower is stiffer and more rigid than the steel tower.

Power is the most important factor, and next to that should come strength, stiffness, and durability.

It has been shown that the steel mills, with their few large sails, have much more power than the wooden mills with their many small sails. (See page 106.) A mill should have as few moving parts as possible, in order that the loss of power by friction shall be small, also the liability to get out of working order be reduced to a minimum. The power of a mill is at best so small that if there is much friction there is little power left to do useful work. The grinder should be on the foot gear and not worked by a belt, and the shafting and cogwheels should not be too heavy. In the large wooden mills the shafting is much too heavy; apparently it is designed on the assumption that the mill will furnish several times more power than it really can. The mill should be carefully erected, the vertical shafting exactly vertical and the horizontal shafting truly horizontal, so that there will be no binding of the parts. Poor workmanship is an important cause of the small power of some mills. Only a skilled workman who understands the business should be employed to erect a windmill.

The mills should be placed at a proper height above surrounding obstructions—at least 30 feet above the highest trees and buildings. This calls for a tower from 50 to 70 feet high. It is better to use a small wheel on a high tower than a large wheel on a low tower. An 8-foot wheel on a 70-foot tower will probably do more work in a given time than a 12-foot wheel on a 30-foot tower with trees and buildings around it. The tower should be firm and rigid, no shaking under a heavy wheel load. Steel towers are in constant vibration under heavy loads.

A mill should govern readily at the proper wind velocity, but this velocity need not be less that 30 miles an hour. A weight appears to be better than a spring for holding the wind wheel in the wind. The tension of a spring can not readily be changed when desired but may gradually lose its tension. (See pages 60 and 110.) There is very great need of an automatic device for changing the load on a pumping mill as the wind velocity changes. The mill should start in a light wind, say 4 to 5 miles an hour, or it will be idle many hours when it should be at work; but in order to do this it must be lightly loaded. In the higher wind velocities, with a light load the mill will do only a small fraction of the work it would do with a much heavier load. The increase in the load should be nearly proportional to the increase in the wind velocity. (See page 113.) Until such a device is invented the load should depend on the wind velocity of the place where the mill is to be used and on the amount of storage.

The pumping mill is ordinarily constructed so that all of the useful work is done on the upstroke of the pump, producing a jerky motion and excessive strain on the working parts. This defect is partly remedied by the use of a large plunger rod, which will force up some of the water on the downstroke. A second remedy is the use of a lever with a heavy weight at one end, the other end being attached to the plunger rod. As the plunger moves down the weight on the end of the lever is raised on the upstroke. The descent of the weight assists the mill in lifting the water. Neither device is satisfactory. A pumping mill working direct stroke makes too many strokes per minute at wind velocities above about 15 miles an hour. The valves ordinarily used for small pumps will not work well if the number of strokes is greater than 30 per minute. The mill should be geared back about 2 to 1 for large mills and about 3 or 4 to 1 for small mills.

INDEX TO PAPERS 41 AND 42.

O

www.ingramcontent.com/pod-product-compliance
Lightning Source LLC
Chambersburg PA
CBHW081126170526
45165CB00008B/2572